IET REN~~EWABLE ENERGY SERIES~~

Distributed Generation

Other volumes in this series:

Distributed Generation

N. Jenkins, J.B. Ekanayake and G. Strbac

The Institution of Engineering and Technology

Published by The Institution of Engineering and Technology, London, United Kingdom

© 2010 The Institution of Engineering and Technology

First published 2010

The Institution of Engineering and Technology
Michael Faraday House
Six Hills Way, Stevenage
Herts, SG1 2AY, United Kingdom

www.theiet.org

British Library Cataloguing in Publication Data
A catalogue record for this product is available from the British Library

ISBN 978-0-86341-958-4 (paperback)
ISBN 978-1-84919-116-6 (PDF)

Typeset in India by MPS Ltd, A Macmillan Company
Printed in the UK by CPI Antony Rowe, Chippenham

Contents

Preface

Public electricity supply was originally developed in the form of local generation feeding local loads, the individual systems being built and operated by independent companies. During the early years, up to around 1930, this proved quite sufficient. However, it was then recognised that an integrated system, planned and operated by a specific organisation, was needed to ensure an electricity supply that was both reasonably secure and economic. This led to large centrally located generators feeding the loads via transmission and distribution systems.

This trend may well have continued but for the need to minimise the environmental impact of energy use (particularly CO_2 emissions) and concern over the security of supply of imported fossil fuels. Consequently, governments and energy planners are now actively developing alternative and cleaner forms of energy production, these being dominated by renewables (e.g. wind, solar, biomass), local CHP plant and the use of waste products. In many countries this transformation of electricity supply is managed through energy markets and privately owned, regulated transmission and distribution systems.

The economics and the location of sustainable energy sources have meant that many of these generators have had to be connected into distribution networks rather than at the transmission level. A full circle has therefore evolved with generation being distributed round the system rather than being located and dispatched centrally.

It is now recognised that the levels of information and control of the state of distribution networks are inadequate for future low-carbon electricity supply systems and the SmartGrid concept has emerged. In addition to much greater use of ICT (Information and Control Technology), a SmartGrid will involve load customers much more in the operation of the power system. The distribution network will change its operation from passive to active, and the distributed generators will be controlled to support the operation of the power system.

Over several decades, models, techniques and application tools have been developed for central generation with traditional transmission and distribution networks and there are many excellent texts that relate to and describe the assessment of such systems. However, some very specific features of distributed generation, namely that large numbers of relatively small generators are distributed around the system, often connected into relatively weak distribution networks, mean that existing techniques and practices have had to be reviewed and updated to take these features into account. The questions raised by large numbers of distributed generators and their control within a deregulated commercial environment (virtual power plants), active network management and microgrids are also addressed.

This book is the outcome of the authors' experience in teaching courses on Distributed Generation to undergraduate and MSc students in the United Kingdom, the United States and Sri Lanka. Many universities are now offering courses on Renewable Energy and how sustainable, low-carbon generation can be integrated effectively into the distribution system. There is a great demand from employers for graduates who have studied such courses but, in almost all countries, a critical shortage of students with a strong education in Electrical Engineering and Power Systems. Hence the book has four tutorial chapters (with examples and questions) to provide fundamental material for those without a strong electrical engineering background. Non-specialists may then benefit from the main body of the text once they have studied the tutorial material.

Finally, no book, including this, could be produced in isolation, and we are indebted to all those colleagues and individuals with whom we have been involved in university, industry and in a number of professional organisations. Particular thanks are due to TNEI Ltd for the use of IPSA and the Manitoba HVDC Centre for the PSCAD/EMTDC simulation programs used in the examples in Chapters 3 and 4. RWE npower renewables and Renewable Energy Systems (RES) generously made available a number of photographs. Also we would like to thank Beishoy Awad for all his help with the drawings.

About the authors

Nick Jenkins was at the University of Manchester (UMIST) from 1992 to 2008. He then moved to Cardiff University where he is now Professor of Renewable Energy. His previous career included 14 years industrial experience, of which five years were spent in developing countries. He is a Fellow of the IET, IEEE and Royal Academy of Engineering. In 2009 and 2010 he was the Shimizu Visiting Professor at Stanford University where he taught a course on 'Distributed Generation and the Grid Integration of Renewables'.

Janaka Ekanayake joined Cardiff University as a Senior Lecturer in June 2008 from the University of Manchester, where he was a Research Fellow. Since 1992 he has been attached to the University of Peradeniya, Sri Lanka and was promoted to Professor of Electrical and Electronic Engineering in 2003. He is a Senior Member of IEEE and a Member of IET. His main research interests include power electronic applications for the power system, and renewable energy generation and its integration. He has published more than 25 papers in refereed journals and has co-authored two books.

Goran Strbac is Professor of Electrical Energy Systems at Imperial College, London. He joined Imperial College in 2005 after 11 years at UMIST and 10 years of industrial and research experience. He led the development of methodologies that underpinned the new UK distribution and transmission network operation, plus design standards and grid codes to include distributed generation. He has conducted impact assessments of large-scale penetration of renewables on the UK generation, transmission and distribution networks for the UK Energy Review and also for the last two Energy White Papers.

Chapter 1
Introduction

Plateau Ardéchois Wind Farm, Rhône Alpes, France (6.8 MW)
850 kW Doubly fed induction generator wind turbines [RES]

1.1 The development of the electrical power system

In the early days of electricity supply, each town or city would have its own small generating station supplying local loads. However, modern electrical power systems have been developed, over the past 70 years, mainly following the arrangement indicated in Figure 1.1. Large central generators of ratings up to 1000 MW and voltages of around 25 kV feed electrical power up through generator transformers to a high voltage interconnected transmission network operating at up to 400 kV in

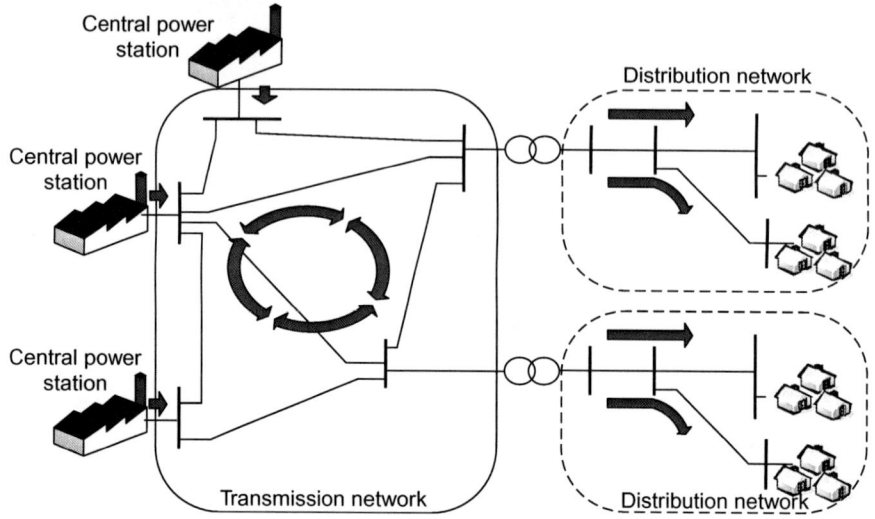

Figure 1.1   Conventional large electric power system

most of Europe and 750 kV in North America and China. The transmission system is used to transport the electrical power, sometimes over considerable distances, and it is then extracted and passed down through a series of distribution transformers to final circuits for delivery to the customers. The transmission and distribution circuits are mainly passive with control of the system provided by a limited number of large central generators [1,2]. From around 1990, there has been a revival of interest in connecting generation to the distribution network and this has come to be known as distributed generation (DG) or the use of distributed energy resources (DER).[1]

The name Distributed Generation can be considered to be synonymous and interchangeable with the terms embedded generation and dispersed generation, which are now falling into disuse. In some countries a strict definition of distributed generation is made, based either on the rating of the plant or on the voltage level to which it is connected. However, these definitions usually follow from particular national, technical documents used to specify aspects of the connection or operation of distributed generation and not from any basic consideration of its impact on the power system. This book is concerned with the fundamental behaviour of generation connected to distribution systems and so a rather broad description of distributed generation is preferred to any particular limits based on plant size, voltage or prime mover type.

Distributed generation may be connected at a number of voltage levels from 120/230 V to 150 kV. Only very small generators may be connected to the lowest voltage networks, but large installations of some hundreds of megawatts are connected to the busbars of high voltage distribution systems (Figure 1.2).

---

[1] The term distributed energy resources includes both distributed generation and controllable load.

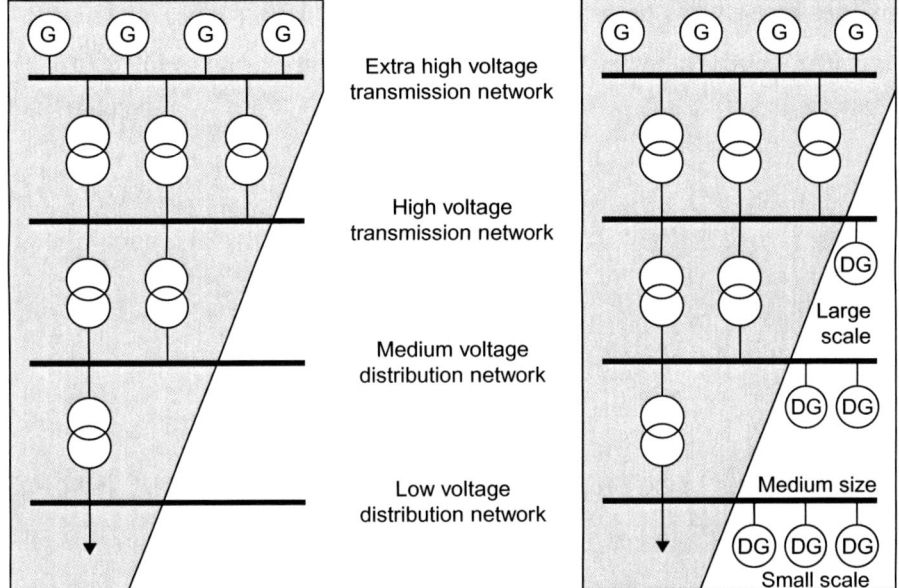

Figure 1.2   Connection of distributed generators

Connecting generation to distribution networks leads to a number of challenges as these circuits were designed to supply loads with power flows from the higher to the lower voltage circuits. Conventional distribution networks are passive with few measurements and very limited active control. They are designed to accommodate all combinations of load with no action by the system operator.

Also, as electrical supply and demand must be balanced on a second-by-second basis, the injection of power from distributed generation requires an equivalent reduction in the output of the large central generators. At present, central generation provides electrical energy but also ancillary services, such as voltage and frequency control, reserve and black-start, which are essential for the operation and stability of the power system. As its use increases, distributed generation will need to provide the ancillary services that are needed to keep the power system functioning with fewer central generators being operated.

1.2   Value of distributed generation and network pricing

For distributed generation to compete successfully with central generation in a competitive environment, network pricing arrangements are critically important. The challenge for realising the value and calculating the impact of generation connected at different levels in the network is illustrated in Figure 1.3 by the value chain from power generation to consumption.

Figure 1.3 shows the price of electricity produced by central generation and sold in the wholesale markets to be around 2–3 p/kWh (the price of wholesale

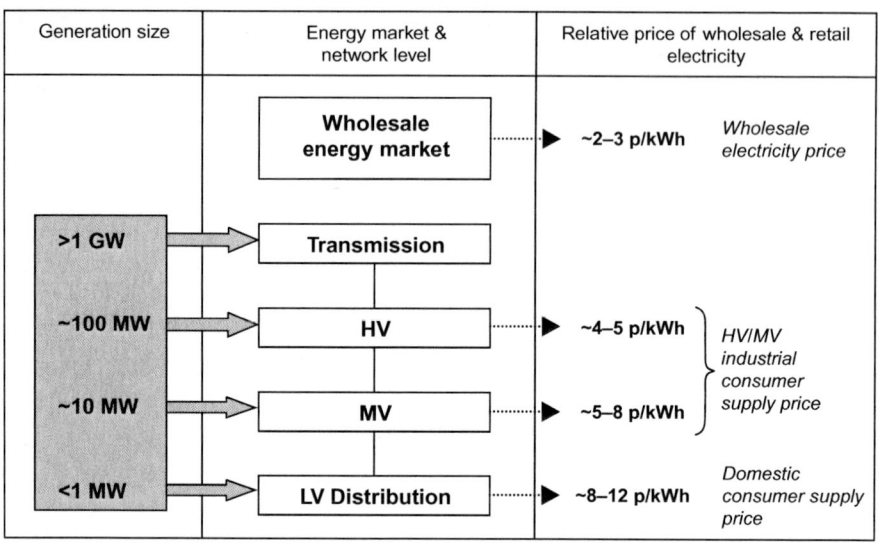

Generation size	Energy market & network level	Relative price of wholesale & retail electricity	
	Wholesale energy market	~2–3 p/kWh	*Wholesale electricity price*
>1 GW	**Transmission**		
~100 MW	**HV**	~4–5 p/kWh	*HV/MV industrial consumer supply price*
~10 MW	**MV**	~5–8 p/kWh	
<1 MW	**LV Distribution**	~8–12 p/kWh	*Domestic consumer supply price*

Figure 1.3   Value chain for electricity from central generation to LV distribution. Prices are for relative comparison only and can also be considered to be in US cents or Euro cents

electricity). By the time electricity reaches the end consumer, the relative 'value' of electricity has increased, and the price, depending on the voltage level in the network, is now 4–12 p/kWh (the retail price of electricity). This increase in the value of electricity up to the point of consumption is driven primarily by the added cost of network transmission and distribution services required to deliver power from central generators to customers elsewhere in the network.

Distributed generation is located closer to the consumer and has fewer requirements for the transport services of the transmission and distribution networks. In essence, distributed generation is delivering power direct to demand at an equivalent value of 4–12 p/kWh, i.e. the costs avoided by not using the network. However, this network cost reduction, generated by the favourable location of some generators, is not fully recognised within the present commercial and regulatory framework. As a consequence, non-conventional generation invariably competes with conventional generation in the wholesale markets at a price (2–3 p/kWh) that may be significantly lower than the true value of electricity delivered from a location close to demand.

The same example is also relevant for the treatment of demand. Customers taking energy from the network at the right location in the network (i.e. close to generation) have less requirement for network services. Yet for most consumers, all the power they receive is priced at the fixed retail rate, occasionally reflecting differences in time of use but never fluctuating in response to location and the distance between generation and the consumer.

The full value of generation in these terms will obviously be dependent on a number of factors, from time of use and location in the system to density of penetration of similar generation and timing of system peaks with output. In some instances renewable and distributed generators may cause more costs than benefit, but the principle of realising the full value of generation still holds. Ignoring these particular features (time of use and location) results in sub-optimal network development because the full and true impact of the user on the network is not represented. Ultimately, this prevents new low-carbon generation (and demand) from competing with incumbent generation and traditional network solutions to guide optimal network development, with the result that the system must resort to increasingly expensive and unnecessary network reinforcement, and sub-optimal network support solutions.

1.3   SmartGrids

Throughout the world, energy policy is developing rapidly with the aims of providing electrical energy supplies that are

1. Low or zero-carbon to reduce the production of greenhouse gases and mitigate climate change.
2. Secure and not dependent on imported fossil fuel.
3. Economic and affordable by industry, commerce and all sections of society.

These objectives of energy policy converge in the use of distributed generation; renewables and cogeneration (combined heat and power, CHP). Recently, the name SmartGrids [3] has become common to describe the future power network that will make extensive use of modern information and communication technologies to support a flexible, secure and cost-effective de-carbonised electrical power system. SmartGrids are intelligently controlled active networks that facilitate the integration of distributed generation into the power system. It is hard to envisage a de-carbonised electrical power system, supplied from renewable energy sources and constant output generation (perhaps nuclear and fossil plant with carbon capture and storage) without greater involvement of the load in its operation. Hence an important aspect of the SmartGrid concept is demand side participation.

Demand side participation is an important potential means of increasing flexibility and controllability in the power system. Controllable loads such as charging electric vehicles and heat pumps with thermal storage will allow increased utilisation of renewable energy. More radically, it is anticipated that the battery energy storage of electric vehicles may be used to inject power into the system at times of high generation cost or to facilitate islanded operation of distribution networks. The role of smart metering and how the customers will wish to control their loads and take part in the operation of the power system is an important topic of investigation with trials being undertaken in many countries at present.

The European Technology Platform for SmartGrids has published a definition of a SmartGrid [3].

"A SmartGrid is an electricity network that can intelligently integrate the actions of all users connected to it – generators, consumers and those that do both – in order to efficiently deliver sustainable, economic and secure electricity supplies.

A SmartGrid employs innovative products and services together with intelligent monitoring, control, communication and self-healing technologies to:

- *better facilitate the connection and operation of generators of all sizes and technologies;*
- *allow electricity consumers to play a part in optimising the operation of the system;*
- *provide consumers with greater information and choice of supply;*
- *significantly reduce the environmental impact of the total electricity supply system and*
- *deliver enhanced levels of reliability and security of supply."*

Smart meters are an important element of a SmartGrid as they have the potential to provide much greater visibility of network power flows and voltages, particularly on low voltage networks where the number of measurements is presently very limited. Details of how the SmartGrids are to be implemented have still to be worked out and will probably vary from country to country. However, there is little doubt of the important consequences of the SmartGrid concept for distributed generators of all types and ratings [4].

1.4   Reasons for distributed generation

The conventional arrangement of a modern large power system (illustrated in Figure 1.1) has a number of advantages. Large generating units can be made efficient and operated with only a relatively small staff. The interconnected high voltage transmission network allows the most efficient generating plant to be dispatched at any time, bulk power to be transported large distances with limited electrical losses and generation reserve to be minimised. The distribution networks can be designed simply for uni-directional flows of power and sized to accommodate customer loads only.

However, particularly in response to climate change, many governments have set ambitious targets to increase the use of renewable energy and to reduce greenhouse gas emissions from electricity generation. Examples of these policy initiatives include the 2007 European Union requirement to provide 20% of all energy used in Europe from renewable sources by 2020 and the California Renewable Portfolio Standard that calls for 33% of electrical energy to be from renewables by the same year. More radically, many climate scientists and policymakers in developed countries consider that an 80% reduction in greenhouse gas emissions by 2050 is necessary if average global temperature rises of more than 2 °C are to be avoided.

The electrical power sector is seen as offering an easier and more immediate opportunity to reduce greenhouse gas emission than, for example, road or air transport and so is likely to bear a large share of any emission reductions. The UK share of the European Union target is only 15% of all energy to come from renewables by 2020 but this translates into some 35% of electrical energy. This target, set in terms of annual electrical energy, will result at times in very large fractions of instantaneous electrical power being supplied from renewables, perhaps up to 60–70%.

Most governments have financial mechanisms to encourage the development of renewable energy generation with opinion divided as to whether feed-in-tariffs, quota requirements (such as the UK Renewables Obligation), carbon trading or carbon taxes provide the most cost-effective approach, particularly for the stimulus of emerging renewable energy technologies. Established technologies include wind power, micro-hydro, solar photovoltaic systems, landfill gas, energy from municipal waste, biomass and geothermal generation. Emerging technologies include tidal stream, wave-power and solar thermal generation.

Renewable energy sources have a much lower energy density than fossil fuels and so the generation plants are smaller and geographically widely spread. For example wind farms must be located in windy areas, while biomass plants are usually of limited size due to the cost of transporting fuel with relatively low energy density. These smaller plants, typically of less than 50–100 MW in capacity, are then connected into the distribution system. It is neither cost-effective nor environmentally acceptable to build dedicated electrical circuits for the collection of this power, and so existing distribution circuits that were designed to supply customers' load are utilised. In many countries the renewable generation plants are not planned by the utility but are developed by entrepreneurs and are not centrally dispatched but generate whenever the energy source is available.

Cogeneration or combined heat and power (CHP) schemes make use of the waste heat of thermal generating plant for either industrial process or space heating and are a well-established way of increasing overall energy efficiency. Transporting the low temperature waste heat from thermal generation plants over long distances is not economic and so it is necessary to locate the CHP plant close to the heat load. This again leads to relatively small generation units, geographically distributed and with their electrical connection made to the distribution network. Although CHP units can, in principle, be centrally dispatched, they tend to be operated in response to the heat requirement or the electrical load of the host installation rather than the needs of the public electricity supply system.

Micro-CHP devices are intended to replace gas heating boilers in domestic houses and, using Stirling or other heat engines, provide both heat and electrical energy for the dwelling. They are operated in response to the demand for heat and hot water within the dwelling and produce modest amounts of electrical energy that is used to offset the consumption within the house. The electrical generator is, of course, connected to the distribution network and can supply electricity back to the network, but financially this is often unattractive with low rates being offered for electricity exported by microgenerators.

The commercial structure of the electricity supply industry plays an important role in the development of distributed generation. In general a deregulated environment and open access to the distribution network is likely to provide greater opportunities for distributed generation although early experience in Denmark provided an interesting counter-example where both wind power and CHP were widely developed within a vertically integrated power system.

1.5   The future development of distributed generation

At present, distributed generation is seen primarily as a means of producing electrical energy and making a limited contribution to the other ancillary services that are required in any power system. Although this is partly due to the technical characteristics of the plant, this restricted role is predominantly caused by the administrative and commercial arrangements under which distributed generation presently operates and is rewarded, i.e. as a source of energy. This is now changing with the transmission connection requirements (the so-called Grid Codes) that specify the performance required from renewable generation connected to transmission networks being applied increasingly to larger distributed generation schemes.

Levels of penetration of distributed and renewable generation in some countries are such that it is already beginning to cause operational problems for the power system. Difficulties have been reported in Denmark, Germany and Spain, all of which have high penetration levels of renewables and distributed generation. This is because, thus far, the emphasis has been on connecting distributed generation to the network in order to accelerate the deployment of all forms of distributed energy resources rather than integrating it into the overall operation of the power system.

The current policies of connecting distributed generation are generally based on a 'fit-and-forget' approach. This is consistent with historic design and operation of passive distribution networks but leads to inefficient and costly investment in distribution infrastructure. Traditionally the distribution network has been designed to allow any combination of load (and distributed generation) to occur simultaneously and still supply electricity to customers with an acceptable power quality. Moreover with passive network operation and simple local generator controls, distributed generation can only displace the energy produced by central generation but cannot displace its capacity as system control and security must continue to be provided by central generation. We are now entering an era where this approach is beginning to restrict the deployment of distributed generation and increase the costs of investment and operation as well as undermine the integrity and security of the power system.

Hence, distributed generation must take over some of the responsibilities from large conventional power plants and provide the flexibility and controllability necessary to support secure system operation. Although transmission system operators have historically been responsible for power system security, the integration of distributed generation will require distribution system operators to develop Active Network Management in order to participate in the provision of system security. This represents a shift from the traditional central control philosophy, presently used to control typically hundreds of generators to a new distributed control paradigm applicable for operation of hundreds of thousands of generators and controllable loads.

Figure 1.4 shows a schematic representation of the capacities (and hence cost) of distribution and transmission networks as well as central generation of today's system and its future development under two alternative scenarios both with increased penetration of distributed energy resources. Business as Usual (BaU) represents traditional system development characterised by centralised control and passive distribution networks as of today. The alternative using SmartGrid concepts and technologies represents the system capacities with distributed generation and the demand side fully integrated into power system operation.

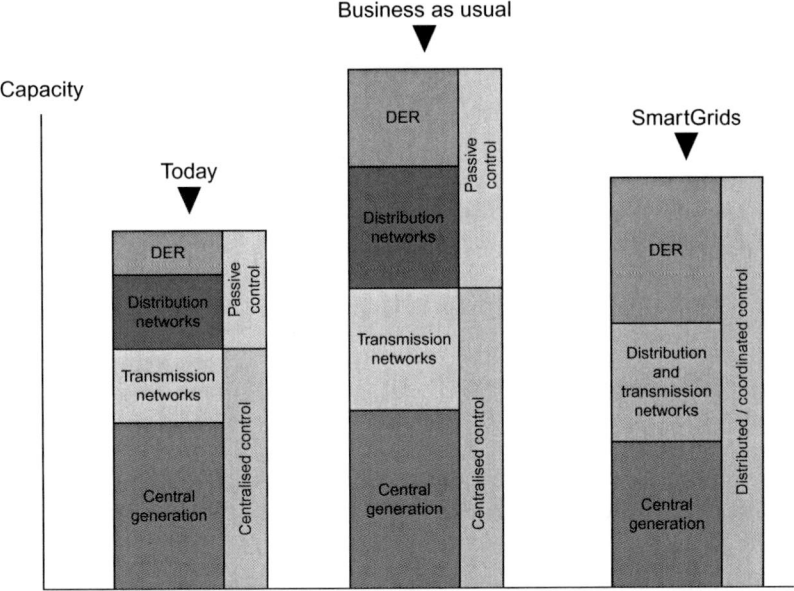

Figure 1.4   Relative levels of system capacity

1.5.1   Business as usual future

Distributed generation will displace energy produced by conventional plant but central generation will continue to be operated for the supply of those ancillary services (e.g. load following, frequency and voltage regulation, reserve) required to maintain the security and integrity of the power system. This leads to large amounts of generation being maintained (a high generation plant margin) and the possible need to operate central generation while distributed generation is deliberately shut down.

The traditional passive operation of the distribution networks and centralised control by the central generators will necessitate increase in capacity of both transmission and distribution networks to accommodate the distributed generation and loads, which are not controlled.

1.5.2   Smart networks

By fully integrating distributed generation and controllable load demand into network operation, these resources will take the responsibility for delivery of some system support services, taking over these from central generation. In this case distributed energy resources (distributed generation and controllable load) will be able to displace not only the energy produced by central generation but also its controllability. This then reduces the capacity of central generation required to be operated and maintained. To achieve this, distribution network operating practice will need to change from passive to active. This will necessitate a shift to a new distributed control paradigm, including significant contribution of the demand side to enhance the control capability of the system.

1.5.3   Benefits of integration

Effective integration of distributed energy resources should bring the following benefits:

- reduced central generation capacity;
- increased utilisation of transmission and distribution network capacity;
- enhanced system security; and
- reduced overall costs and CO_2 emissions.

The expected total system costs of these two futures are represented in Figure 1.5. In the short term, the change in the control and operating philosophy of distribution networks is likely to increase costs over Business as Usual (BaU). Additional costs are required to cover research, development and deployment of new technologies and the required information and communication infrastructure. However, full integration of distributed generation and responsive demand using SmartGrid concepts and technologies will deliver benefits over the longer term.

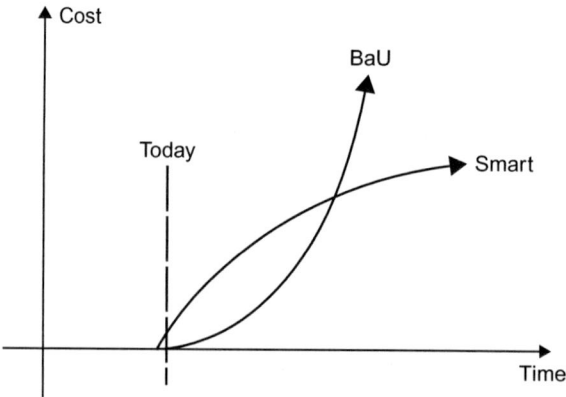

Figure 1.5   Trajectories of expected future system costs

The future based on the concepts of SmartGrids will require a more sophisticated commercial structure to enable individual participants, distributed generators and demand customers or their agents to trade not only energy but also various ancillary services. The development of a market with hundreds of thousands of active participants will be a major challenge.

1.6   Distributed generation and the distribution system

Traditionally, distribution systems have been designed to accept power from the transmission network and to distribute it to customers. Thus the flow of both real power (P) and reactive power (Q) has been from the higher to the lower voltage levels. This is shown schematically in Figure 1.6 and, even with interconnected distribution systems, the behaviour of such networks is well understood and the procedures for both design and operation long established.

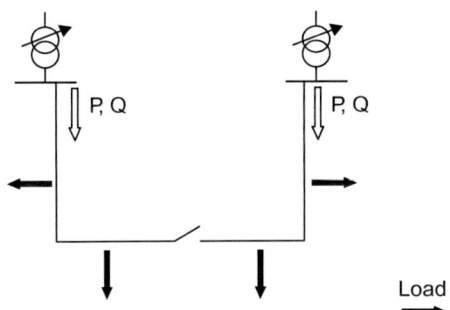

Figure 1.6   Conventional distribution system

However, with significant penetration of distributed generation the power flows may become reversed and the distribution network is no longer a passive circuit supplying loads but an active system with power flows and voltages determined by the generation as well as the loads. This is shown schematically in Figure 1.7. For example the combined heat and power (CHP) scheme with the synchronous generator (S) will export real power when the electrical load of the premises falls below the output of the generator but may absorb or export reactive power depending on the setting of the excitation system of the generator. The fixed-speed wind turbine will export real power but is likely to absorb reactive power as its induction (sometimes known as asynchronous) generator (A) requires a source of reactive power to operate. The voltage source converter of the photovoltaic (PV) system will allow export of real power at a set power factor but may introduce harmonic currents, as indicated in Figure 1.7. Thus the power flows through the circuits may be in either direction depending on the relative magnitudes of the real and reactive network loads compared to the generator outputs and any losses in the network.

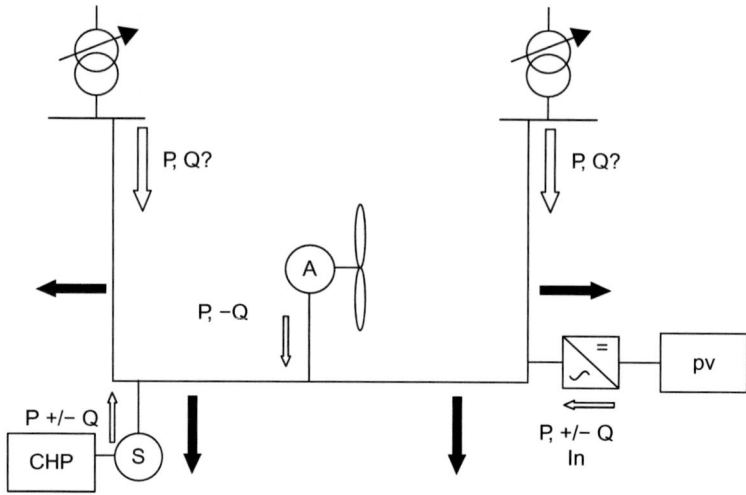

Figure 1.7   Distribution system with distributed generation

The change in real and reactive power flows caused by distributed generation has important technical and economic implications for the power system. In the early years of distributed generation, most attention was paid to the immediate technical issues of connecting and operating generation on a distribution system and most countries developed standards and practices to deal with these [5,6]. In general, the approach adopted was to ensure that any distributed generation did not reduce the quality of supply voltage offered to other customers and to consider the generators as negative load. The philosophy was of fit-and-forget where the distribution system was designed and constructed so that it functioned correctly for all combinations of generation and load with no active control actions, of the generators, loads or networks, being taken. This approach, with the distribution system configuration determined at the planning stage and with no operational control, is in contrast to that adopted on the transmission system where active control of the central generators by the system operator in real time is necessary.

1.7   Technical impacts of generation on the distribution system

1.7.1   Network voltage changes

Every distribution network operator has an obligation to supply its customers at a voltage within specified limits (typically around ±5% of nominal). This requirement often determines the design and capital cost of the distribution circuits and so, over the years, techniques have been developed to make the maximum use of distribution circuits to supply customers within the required voltages.

The voltage profile of a radial distribution feeder is shown in Figure 1.8 with the key volt drops identified:

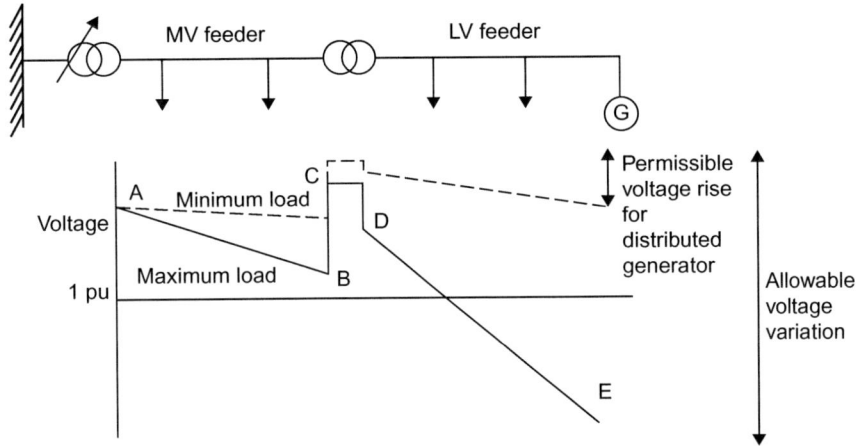

Figure 1.8   Voltage variation down a radial feeder [after Reference 7]

- A: voltage held constant by tap-changer of distribution transformer
- A–B: voltage drop due to load on medium voltage (MV) feeder
- B–C: voltage boost due to taps of MV/LV transformer
- C–D: voltage drop in MV/LV transformer
- D–E: voltage drop in LV feeder

The precise voltage levels used differ from country to country, but the principle of operation of distribution using radial feeders remains the same. Table 1.1 shows the normal voltage levels used.

Table 1.1   Voltage levels used in distribution circuits

	Definition	Typical UK voltages
Low voltage (LV)	LV<1 kV	230 (1 phase)/400 (3 phase) V
Medium voltage (MV)	1 kV<MV<50 kV	33 kV, 11 kV
High voltage (HV)	50 kV<HV<150 kV	132 kV

Figure 1.8 shows that the ratio of the MV/LV transformer has been adjusted at its installation, using off-circuit taps so that at times of maximum load the most remote customer will receive acceptable voltage. During minimum load the voltage received by all customers is just below the maximum allowed. If a distributed generator is now connected to the end of the circuit, then the flows in the circuit will change and hence the voltage profile. The most onerous case is likely to be

when the customer load on the network is at a minimum and the output of the distributed generator must flow back to the source [8].

For a lightly loaded distribution network the approximate voltage rise (ΔV) caused by a generator exporting real and reactive power is given by (1.1):[2]

$$\Delta V = \frac{PR + XQ}{V} \qquad (1.1)$$

where P = active power output of the generator, Q = reactive power output of the generator, R = resistance of the circuit, X = inductive reactance of the circuit and V = nominal voltage of the circuit.

In some cases, the voltage rise can be limited by reversing the flow of reactive power (Q). This can be achieved either by using an induction generator, under-exciting a synchronous machine or operating an inverter so as to absorb reactive power. Reversing the reactive power flow can be effective on medium voltage overhead circuits which tend to have a higher ratio of X/R of their impedance. However, on low voltage cable distribution circuits the dominant effect is that of the real power (P) and the network resistance (R). Only very small distributed generators may generally be connected out on low voltage networks.

For larger generators a point of connection of the generator is required either at the low voltage busbars of the MV/LV transformer or, for even larger plants, directly to a medium voltage or high voltage circuit. In some countries, simple design rules have been used to give an indication of the maximum capacity of distributed generation which may be connected at different points of distribution system. These simple rules tend to be rather restrictive and more detailed calculations can often show that more generation can be connected with no difficulties. Table 1.2 shows some of the rules that have been used.

Table 1.2   Design rules sometimes used for an indication if a distributed generator may be connected

Network location	Maximum capacity of distributed generator
Out on 400 V network	50 kVA
At 400 V busbars	200–250 kVA
Out on 11 kV or 11.5 kV network	2–3 MVA
At 11 kV or 11.5 kV busbars	8 MVA
On 15 kV or 20 kV network and busbars	6.5–10 MVA
Out on 63 kV or 90 kV network	10–40 MVA

An alternative simple approach to deciding if a generator may be connected is to require that the three-phase short-circuit level (fault level) at the point of connection, before the generator is connected, is a minimum multiple of the distributed

---

[2] See Section 3.3.1 for the derivation of this equation.

generator power rating. Multiples as high as 20 or 25 have been required for wind turbines/wind farms in some countries, but again these simple approaches are very conservative. Large wind farms using fixed-speed induction generators have been successfully operated on distribution networks with a ratio of network fault level to wind farm rated capacity as low as 6.

Some distribution companies use controls of the on-load tap changers at the distribution transformers, based on a current signal compounded with the voltage measurement. One technique is that of line drop compensation [7] and, as this relies on an assumed power factor of the load, the introduction of distributed generation and the subsequent change in power factor may lead to incorrect operation if the distributed generation on the circuit is large compared to the customer load.

1.7.2   Increase in network fault levels

Many types of larger distributed generation plant use directly connected rotating machines and these will contribute to the network fault levels. Both induction and synchronous generators will increase the fault level of the distribution system although their behaviour under sustained fault conditions differs.

In urban areas where the existing fault level approaches the rating of the switchgear, this increase in fault level can be a serious impediment to the development of distributed generation schemes. Increasing the short-circuit rating of distribution network switchgear and cables can be extremely expensive and difficult particularly in congested city substations and cable routes. The fault level contribution of a distributed generator may be reduced by introducing impedance between the generator and the network, with a transformer or a reactor, but at the expense of increased losses and wider voltage variations at the generator. In some countries fuse-type fault current limiters are used to limit the fault-level contribution of distributed generation plant and there is also continued interest in the development of superconducting fault current limiters.

1.7.3   Power quality

Two aspects of power quality [9] are usually considered to be important with distributed generation: (1) transient voltage variations and (2) harmonic distortion of the network voltage. Depending on the particular circumstance, distributed generation plant can either decrease or increase the quality of the voltage received by other users of the distribution network.

Distributed generation plant can cause transient voltage variations on the network if relatively large current changes during connection and disconnection of the generator are allowed. The magnitude of the current transients can, to a large extent, be limited by careful design of the distributed generation plant although for single, directly connected induction generators on weak systems the transient voltage variations caused may be the limitation on their use rather than steady-state voltage rise. Synchronous generators can be connected to the network with negligible disturbance if synchronised correctly and anti-parallel soft-start units can be used to limit the magnetising inrush of induction generators to less than rated current. However, disconnection of the generators when operating at full output may lead to significant, if infrequent, voltage drops. Also, some forms of prime mover (e.g. fixed speed wind turbines) may cause cyclic variations in the generator output current, which can lead to so-called flicker nuisance if not adequately controlled.

Conversely, however, the addition of rotating distributed generation plant acts to raise the distribution network fault level. Once the generation is connected and the short-circuit level increased, any disturbances caused by other customers loads, or even remote faults, will result in smaller voltage variations and hence improved power quality. It is interesting to note that one conventional approach to improving the power quality in sensitive, high value manufacturing plants is to install local generation.

Similarly, incorrectly designed or specified distributed generation plant, with power electronic interfaces to the network, may inject harmonic currents which can lead to unacceptable network voltage distortion. The large capacitance of extensive cable networks or shunt power factor correction capacitors may combine with the reactance of transformers or generators to create resonances close to the harmonic frequencies produced by the power electronic interfaces.

The voltages of rural MV networks are frequently unbalanced due to the connection of single phase transformers. An induction generator has very low impedance to unbalanced voltages and will tend to draw large unbalanced currents and hence balance the network voltages at the expense of increased currents in the generator and consequent heating.

1.7.4 Protection

A number of different aspects of distributed generator protection can be identified as follows:

• Protection of the distributed generator from internal faults
• Protection of the faulted distribution network from fault currents supplied by the distributed generator
• Anti-islanding or loss-of-mains protection
• Impact of distributed generation on existing distribution system protection

Protecting the distributed generator from internal faults is usually fairly straightforward. Fault current flowing from the distribution network is used to detect the fault and techniques used to protect any large motor or power electronic converter are generally appropriate. In rural areas with limited electrical demand, a common problem is ensuring that there will be adequate fault current from the network to ensure rapid operation of the relays or fuses.

Protection of the faulted distribution network from fault current from the distributed generators is often more difficult. Induction generators cannot supply sustained fault current to a three-phase balanced fault, and their sustained contribution to asymmetrical faults is limited. Small synchronous generators require sophisticated exciters and field forcing circuits if they are to provide sustained fault current significantly above their full load current. Insulated gate bipolar transistor (IGBT) voltage source converters often can only provide close to their continuously rated current into a fault. Thus, it is usual to rely on the distribution protection and fault current from the network to clear any distribution circuit fault and hence isolate the distributed generation plant. The distributed generator is then tripped on over/under voltage, over/under-voltage protection or loss-of-mains protection.

Loss-of-mains protection is an important issue in a number of countries, particularly where auto-reclose is used on the distribution circuits. For a variety of reasons, both technical and administrative, the prolonged operation of a power island fed from the distributed generation but not connected to the main distribution network is considered unacceptable. Thus a relay is required which will detect when the distributed generator, and perhaps a surrounding part of the network, has become islanded and will then trip the generator. This relay must work within the dead time of any auto-reclose scheme if out-of-phase reconnection is to be avoided. Although a number of techniques are used, including rate-of-change-of-frequency (ROCOF) and voltage vector shift, these are prone to maloperation if set sensitively to detect islanding rapidly.

The neutral grounding of the generator is a related issue as in a number of countries it is considered unacceptable to operate an ungrounded power system and so care is required to ensure a neutral connection is maintained.

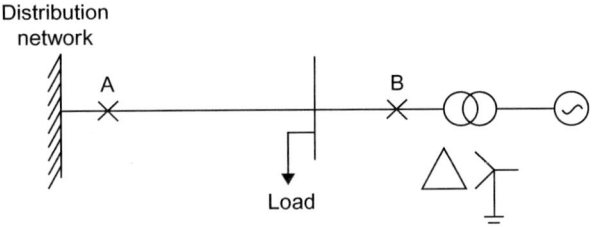

Figure 1.9   Illustration of the 'islanding' problem

The loss-of-mains or islanding problem is illustrated in Figure 1.9. If circuit breaker A opens, perhaps on a transient fault, there may well be insufficient fault current from the generator to operate the protection of circuit breaker B. In this case the generator may be able to continue to supply the load. If the output of the generator is able to match the real and reactive power demand of the load closely, then there will be no change in either the frequency or voltage of the islanded section of the network and so a sequential tripping scheme based on under/over frequency or voltage will fail to operate. Thus, it is very difficult to detect reliably that circuit breaker A has opened using only local measurements at B. In the limit, if there is no current flowing through A (the generator is supplying the entire load), then the network conditions at B are unaffected whether A is open or closed. It may also be seen that since the load is being fed through the delta winding of the transformer, there is no neutral earth on that section of the network.

Finally, distributed generation may affect the operation of existing distribution network by providing flows of fault current that were not anticipated when the protection was originally designed. The fault contribution from a distributed generator can support the network voltage and lead to distance relays under-reaching.

1.7.5   Stability and fault ride through

For distributed generation schemes, whose object is to produce energy considerations of generator transient stability have tended not to be of great significance. If a fault occurs somewhere on the distribution network to depress the network voltage, the distributed generator will over-speed and trip on its internal protection. Then all that is lost is the output of a short period of generation. The control scheme of the distributed generator will then wait for the network conditions to be restored and restart automatically after some minutes. Of course if the generation scheme is intended mainly as a provider of steam for a critical process, then more care is required to try to ensure that the generator does not trip for remote network faults. However, as the inertia of distributed generator is often low, and the tripping time of distribution protection long, it may not be possible to ensure stability for all faults on the distribution network.

In contrast, if a distributed generator is viewed as providing support for the power system, then its transient stability becomes of considerable importance. Both voltage and/or angle stability may be significant depending on the generator type used. A particular problem in some countries is nuisance tripping of the loss of mains relays based on measurements of rate of change of frequency (df/dt). These are set sensitively to detect islanding, but in the event of a major system disturbance, for example, loss of a large central generator, they maloperate and trip large amounts of distributed generation. The effect of this is, of course, to depress the system frequency further. This requirement to stay connected to the power system and support it during network faults is a key attribute required of transmission-connected renewable generation and referred in the Grid Codes as fault ride through.

Synchronous generators will pole-slip during transient instability, but when induction generators over-speed they draw very large reactive currents, which then depress the network voltage further and lead to voltage instability. The steady-state stability limit of induction generators can also limit their application on weak distribution networks as a high source impedance, or low network short-circuit level, can reduce their peak torque to such an extent that they cannot operate at rated output.

The restoration, after an outage, of a section of the distribution network with significant distributed generation (black-start) also requires care. If the circuit was relying on the distributed generation to support its load then, once the circuit is restored, the load will demand power before the generation can be reconnected. This is, of course, a common problem faced by operators of central generation/ transmission networks but is encountered less often in distribution systems.

1.8   Economic impact of distributed generation on the distribution system

Distributed generation alters the power flows in the network and so will alter network losses. If a small distributed generator is located close to a large load, then the network losses will be reduced as both real and reactive power can be supplied to the load from the adjacent generator. Conversely, if a large distributed generator is located far away from network loads, then it is likely to increase losses on the distribution system. A further complication arises due to the changing value of electrical energy as the network load increases. In general there is a correlation between high load on the distribution network and the operation of expensive central generation plant. Thus, any distributed generator that can operate in this period and reduce distribution network losses will make a significant impact on the costs of operating the network.

At present, distributed generation generally takes no part in the voltage control of distribution networks. Thus, in Great Britain, distributed generators will generally choose to operate at unity power factor in order to minimise their electrical losses and avoid any charges for reactive power consumption, irrespective of the needs of the distribution network. Some years ago in Denmark an alternative approach was developed with distributed CHP schemes operating at different power factors according to the time of day. During periods of peak loads reactive power was exported to the network, while during low network load they operated at unity power factor.

Distributed generation can also be used as a substitute for distribution network capacity. Clearly, distributed generators cannot substitute for radial feeders, as islanded operation is not generally acceptable, and network extensions may be required to collect power from isolated renewable energy schemes. However, most high voltage distribution circuits are duplicated or meshed and distributed generation can reduce the requirement for these assets.

1.9   Impact of distributed generation on the transmission system

In a similar manner to the distribution system, distributed generation will alter the flows in the transmission system. Hence transmission losses will be altered, generally reduced, while in a highly meshed transmission network it may be demonstrated that reduced flows lead to a lower requirement for assets. In Great Britain, the charges for use of the transmission network are presently evaluated based on a measurement of peak demand at the transformer linking the transmission and distribution networks. When distributed generation plant can be shown to be operating during the periods of peak demand, then it is clearly reducing the charges for use of the transmission network.

1.10   Impact of distributed generation on central generation

The main impact of distributed generation on central generation has been to reduce the mean level of the power output of the central generators but, often, to increase its variance. In a large electrical power system, consumer demand can be predicted accurately by the generator dispatching authority. Distributed generation will introduce additional uncertainty in these estimates and so may require additional reserve plant. It is now conventional to predict the output of wind farms, by forecasting wind speeds, and distributed CHP plants, by forecasting heat demand. Over a long period, forecasting the output of wind farms shows significant benefit and so is useful for energy trading. These forecasts are, however, less useful for dispatching conventional generators where a very high level of reliability of the forecasts is required.

As distributed generation is added to the system, its output power must displace an equal output of central generators in order to maintain the overall load/generation balance. With limited output of distributed generators, the effect is to deload the central generators but maintain them and their controllable output, on the power system. However, as more and more distributed generation is added it becomes necessary to disconnect central generation with consequent loss of controllability and frequency regulation. A similar effect will be encountered with reduction in reactive power capability as central generators are displaced and difficulties maintaining the voltage profile of the transmission network may be anticipated.

References

1. Blume S.W. *Electric Power System Basics for the Nonelectrical Professional*. IEEE Press; 2007.
2. Von Meier A. *Electric Power Systems: A Conceptual Introduction*. Hoboken, NJ: John Wiley and Sons; 2006.
3. Commission of the European, Union Smart Grids Technology Platform. European Technology Platform for the Electricity Networks of the Future. Available from URL http://www.smartgrids.eu/ [Accessed February 2010].
4. Department of Energy and Climate Change (DECC). Developing a UK Smart Grid. ENSG Vision Statement. Available from URL http://www.ensg.gov.uk/assets/ensg_smart_grid_wg_smart_grid_vision_final_issue_1.pdf [Accessed February 2010].
5. IEEE 1547. IEEE Standard for Interconnecting Distributed Resources with Electric Power Systems; 2003.
6. Electricity Network Association Engineering Recommendation G59/1. Recommendations for the Connection of Embedded Generation Plant to the Public Electrical Suppliers Distribution Systems; 1991.
7. Lakervi E., Holmes E.J. *Electricity Distribution Network Design*. London: Peter Peregrinus for the IEE; 1989.
8. Masters C.L. 'Voltage rise: The big issue when connecting embedded generation to long 11 kV overhead lines'. *IET Power Engineering Journal*. 2002; **16**(1):512.
9. Dugan S., McGranaghan M.F., Beaty H.W. *Electrical Power Systems Quality*. New York: McGraw Hill; 1996.

Chapter 2
Distributed generation plant

Nine Canyon Wind Farm, Tri-Cities, Washington USA (100 MW)
Attaching the blade assembly of a 1.3 MW, active stall regulated, fixed speed
induction generator wind turbine [RES]

2.1 Introduction

With the increasing efforts worldwide to de-carbonise energy supply, a wide variety of generating plant types is being connected to electrical distribution networks.
Examples include the well-established technologies of combined heat and power
(CHP), wind turbines and photovoltaic systems. In addition there are many newer
technologies, such as fuel cells, solar thermal, micro-CHP and the marine renewable technologies as well as flywheel and flow battery storage that are at various
stages of demonstrating their commercial viability.

In deregulated electricity supply systems the owners of the distributed generation) plant (who will not be the distribution utility in most cases) will respond to pricing and other commercial signals to determine whether to invest in such plant and then how to operate it. As distributed generation displaces large central generating plant, it will increasingly take over the ancillary services (e.g. voltage and frequency control) that are necessary for the operation of the power system and so distributed generation will develop from being a source only of energy to being an integral contributor to the supply of electrical power.

2.2   Combined heat and power plants

Combined heat and power (sometimes known as 'cogeneration' or 'total energy') is a significant source of generation connected to distribution networks. Combined heat and power (CHP) is the simultaneous production of electrical power and useful heat. Generally the electrical power is consumed inside the host premises or industrial plant of the CHP facility although any surplus or deficit is exchanged with the public distribution system. The heat generated is either used for industrial processes and/or for space heating inside the host premises or alternatively is transported to the surrounding area for district heating.

A typical industrial CHP scheme might achieve an overall efficiency of 67%, electrical efficiency of 23% and heat efficiency of 44%. This would lead to some 35% reduction in primary energy use compared to electrical generation from single-cycle central power stations and heat only boilers. This corresponds to a reduction in CO_2 emissions of over 30% in comparison with large coal-fired stations and some 10% in comparison with central combined-cycle gas turbine plant. Although the calculations are involved, it is estimated that UK savings in emissions in 2008 due to CHP were 10.8 Mt/CO_2 or 1.98 Mt/CO_2 per 1000 MWe of CHP capacity [1].

The use of CHP for district heating is limited in the United Kingdom although small schemes exist in some cities. However, in Northern Europe, for example Denmark, Sweden and Finland, district heating is common in many large towns and cities with the heat supplied, at water temperatures in the range 80–150 °C, either from CHP plant or heat only boilers [2]. Denmark has extended the use of CHP into rural areas with the installation of smaller CHP schemes in villages and small towns using either back-pressure steam turbines, fed from biomass in some cases, or reciprocating engines powered by natural gas [3].

Table 2.1 lists the various technologies used in CHP plants. Back-pressure steam turbines exhaust steam at greater than atmospheric pressure either directly to an industrial process or to a heat exchanger. The higher the back pressure the more energy there is in the exhausted steam and so the less electrical power that is produced. The back-pressure steam turbines in the United Kingdom had an average heat to power ratio of 5.4:1 and so, once the site electrical load has been met, any export of electrical power will be small. Figure 2.1 is a simplified diagram of a CHP scheme using a back-pressure steam turbine. All the steam passes through the turbine that drives a synchronous generator, usually operating at 3000 rpm. After the turbine the steam, at a pressure typically in the range 0.12–4 MPa and a temperature

Table 2.1 A summary of CHP schemes in the United Kingdom in 2008 [1]

Main prime mover	Average electrical efficiency (%)	Average heat efficiency (%)	Average heat/power ratio
Back-pressure steam turbine	12	63	5.4:1
Pass-out condensing steam turbine	14	44	3.2:1
Gas turbine	21	49	2.3:1
Combined cycle	26	41	1.6:1
Reciprocating engine	26	42	1.3:1
All schemes	23	44	1.9:1

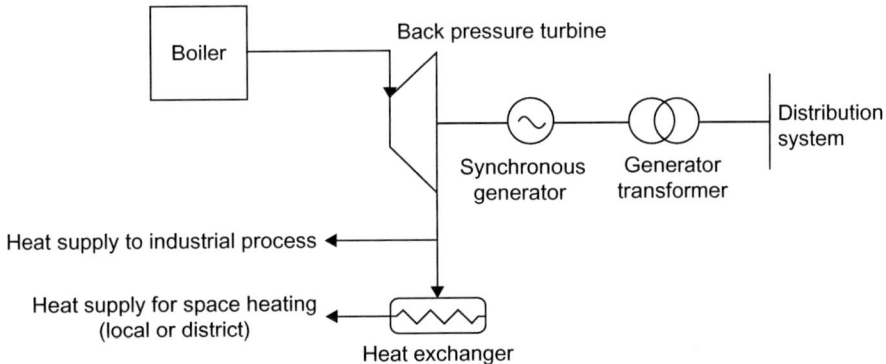

Figure 2.1 CHP scheme using a back-pressure steam turbine

between 200 and 300 °C [4] depending on its use, is passed to the industrial process or through a heat exchanger for use in space heating.

In contrast, in a pass-out (or extraction) condensing steam turbine (Figure 2.2), some steam is extracted at an intermediate pressure for the supply of useful heat with the remainder being fully condensed. This arrangement allows a wide range of

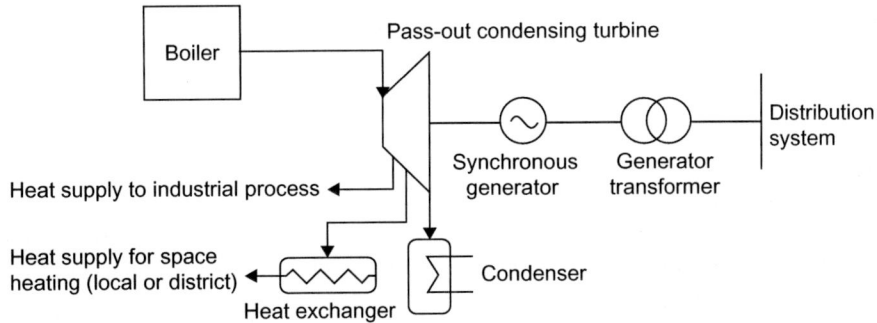

Figure 2.2 CHP scheme using a pass-out condensing steam turbine

heat/power ratios. In Denmark all the large town and city district-heating schemes (150–350 MWe) use this type of unit which, taking into account the capacity that is reduced for heat load, can be dispatched in response to the electrical power demand of the public utilities.

Figure 2.3 shows how the waste heat from a gas turbine may be used. Gas turbines using either natural gas or distillate oil liquid fuel are available in ratings from less than 1 MWe to more than 100 MWe although at the lower ratings internal combustion reciprocating engines may be preferred for CHP schemes. In some industrial installations, provision is made for supplementary firing of the waste heat boiler in order to ensure availability of useful heat when the gas turbine is not operating or to increase the heat/power ratio. The exhaust gas temperature of a gas turbine can be as high as 500–600 °C and so, with the large units, there is the potential to increase the generation of electrical power by adding a steam turbine and to create a combined-cycle plant (Figure 2.4). Steam is raised, using the exhaust gases of the gas turbine passed through a waste heat boiler, which is fed to either a back-pressure or pass-out condensing steam turbine. Useful heat is then recovered from the steam turbine.

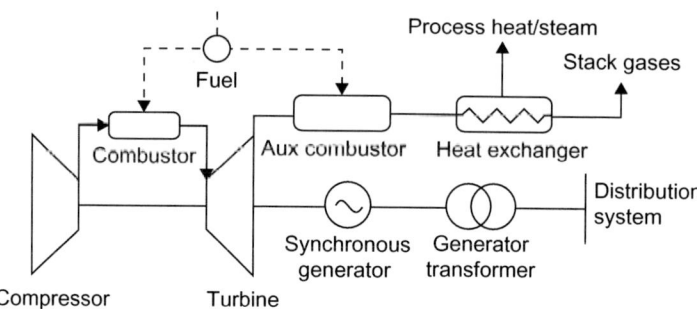

Figure 2.3   CHP scheme using a gas turbine with waste heat recovery

In CHP schemes using a combined cycle, some 67% of the energy in the fuel is transformed into electrical power or useful heat but with a lower heat/power ratio than single-cycle gas turbines (see Table 2.1). Combined-cycle CHP plants, because of their complexity and capital cost, tend to be suitable for large electric and heating loads such as the integrated energy supply to a town or large industrial plant [5].

Although in the United Kingdom installed CHP capacity is dominated by plants with ratings greater than 10 MWe, there are presently over 1000 CHP installations with ratings less than 1 MWe and some 500 with ratings less than 100 kWe. These plants of less than 100 kWe are typically skid-mounted units consisting of a reciprocating four-stroke engine driving a three-phase synchronous, or in some cases, asynchronous generator, with a heat recovery system to extract heat from the exhaust gases, the cooling water and the lubrication oil [6]. For larger engines (i.e. approaching 500 kW), it becomes economic to pass the exhaust gas, which may be at up to 350–400 °C, to a steam-raising waste heat boiler. The heat available from the cooling jacket and the lubrication oil is typically at 70–80 °C. The fuel used is

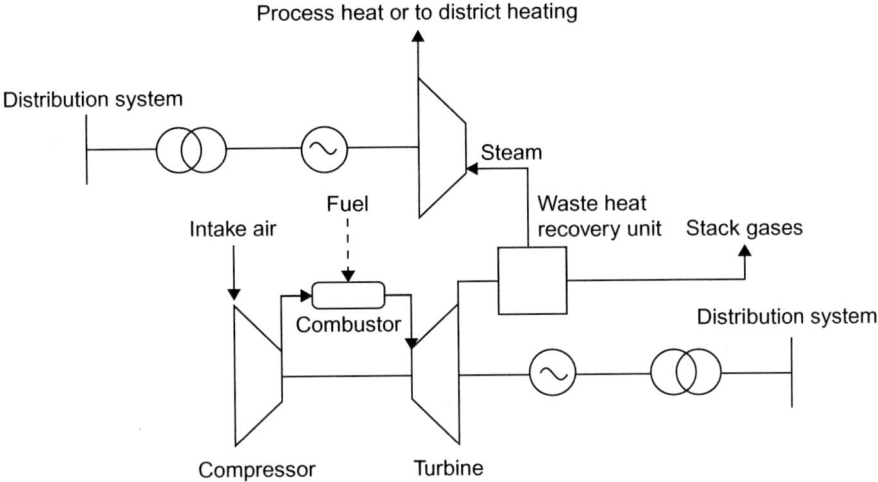

Figure 2.4   CHP scheme with combined-cycle generator (back-pressure steam turbine)

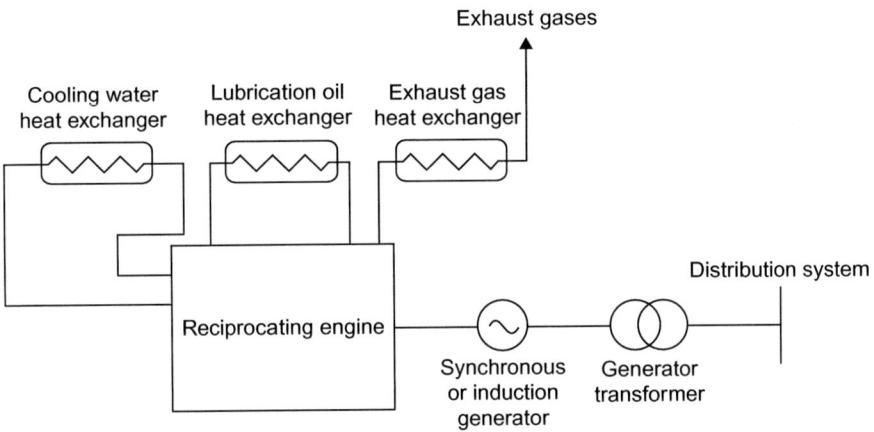

Figure 2.5   CHP scheme using a reciprocating engine with heat recovery

usually natural gas, sometimes with a small addition of fuel oil to aid combustion, or, in some cases, digester gas from sewage treatment plants. Typical applications include leisure centres, hotels, hospitals, academic establishment and industrial processes. Landfill gas sites tend to be too far away from a suitable heat load and so the engines, fuelled by the landfill gas, are operated as electrical generating sets only and not in CHP mode.

The economic case for a CHP scheme depends both on the heat and electrical power load of the host site and critically on the so-called 'spark spread', the difference between the price of electricity and of the gas required to generate it. A wide spark spread makes CHP schemes more attractive. Commercial viability also depends on the rate received for exporting electrical power to the utility and this can be very low for small installations.

CHP units are typically controlled, or dispatched, to meet the energy needs of the host site and not to export electrical power to the utility distribution system [7]. It is common for CHP units to be controlled to meet a heat load, and in district-heating schemes, the heat output is often controlled as a function of ambient temperature. Alternatively, the units can be controlled to meet the electrical load of the host site and any deficit in the heat requirement is met from an auxiliary source. Finally, the units may be run to supply both heat and electricity to the site in an optimal manner, but this is likely to require a more sophisticated control system that is able to compensate for the changes in heat and electrical load as well as in the performance of the CHP plant over time.

Although CHP schemes are conventionally designed and operated to meet the energy needs of the host site, or a district-heating load, this is a commercial/economic choice rather than being due to any fundamental limitation of the technology. As commercial and administrative conditions change, perhaps in response to the policy drivers described in Chapter 1, it may be that CHP plants will start to play a more active role in supplying electrical energy and ancillary services to the electricity distribution system.

CHP units for individual houses or small commercial properties, so-called micro-CHP, are also commercially available. These typically use small internal combustion or Stirling cycle engines coupled to simple induction generators although variable speed generators connected through a power electronic interface are also used. Some designs use the reciprocating motion of the piston directly to drive a linear electrical generator. Prototypes also exist of domestic fuel-cell CHP units, which produce direct current and so require a power electronic interface. In general micro-CHP units are more suited to larger dwellings with a significant heat demand and are less attractive for small, very well insulated houses.

Heat energy is much easier and cheaper to store than electrical energy and so heat stores can be used to increase the flexibility of the operation of CHP units. In Denmark, the direct relationship between heat and electrical power production from reciprocating engine CHP units was a concern, as it would have imposed significant additional load variations on the larger power generating units when the dispersed CHP units responded to the varying demand for district heating. Therefore, large heat stores were constructed for each district-heating scheme to accommodate approximately 10 hours of maximum heat production [3]. One benefit of the heat stores is that they allow the CHP units to be run for reduced periods but at rated output, and hence maximum efficiency. Also, the times of operation can be chosen to be at periods of maximum electrical demand and so the CHP generation can respond to the needs of the electrical power network and receive a higher price for their electricity. An additional advantage of large heat stores is that, with electric heaters, they can use excess energy from intermittent renewables (e.g. wind power) when there is a surplus of electrical energy and its value is low.

2.3   Renewable energy generation

The location of distributed generators used for CHP is fixed by the position of the heat load, and their operation is generally controlled in response to the energy demands of the host site or of a district-heating scheme. Similarly, the siting of generators using renewable energy sources is determined by the location of the renewable energy resource, and their output follows the availability of the resource. Certainly, the choice of location of any generation plant is limited by the local environmental impact it may have, but it is obvious that, for example, the position of a small-scale hydro generating station (and hence the possibilities for connecting it to the distribution network) must be determined by the location of the hydro resource. Unless the renewable energy resource can be stored (e.g. as potential energy of water behind dams or by storing biomass), the generator will operate when the energy is available. In general it is not cost-effective to provide large energy stores for small renewable energy schemes and so their output varies as the resource becomes available. This is an important difference compared with distributed generators that use fossil fuel which, because of its much higher energy density, can be stored economically.

2.3.1   Small-scale hydro generation

Hydro generation is a relatively mature technology, and many of the good sites in Europe and North America have been developed. The operation of small- and medium-sized hydro generating units in parallel with the distribution system is well understood [8,9] although there is scope for innovation through the use of variable speed generation systems. Those hydro schemes without significant water storage capacity may experience large variations in available water flow. Hence their output varies with rainfall, particularly if the catchment is on rocky or shallow soil with a lack of vegetation cover and steep, short streams. Uneven rainfall will then lead to variable flow in the rivers, and it is interesting to note that the capacity factor for hydro generation in the United Kingdom for 2008 was only some 35%.[1]

Figure 2.6 shows the average daily flows of two rivers with catchment areas of very different characteristics leading to very different hydrological resources. Figure 2.6(a) shows the flow in a river with a rocky catchment and little water storage capacity. In contrast, Figure 2.6(b) shows the much smoother flow from a sandy catchment.

It is conventional to express the resource as a flow duration curve (FDC), which shows the percentage time that a given flow is equalled or exceeded (Figure 2.7) [10]. Although an FDC gives useful information about the annual energy yield that may be expected from a given hydro resource, it provides no information as to how

---

[1] Capacity (or load) factor is the ratio of annual energy generated to that which would be generated with the plant operating at rated output all year. UK onshore wind farms have a capacity factor of around 27%.

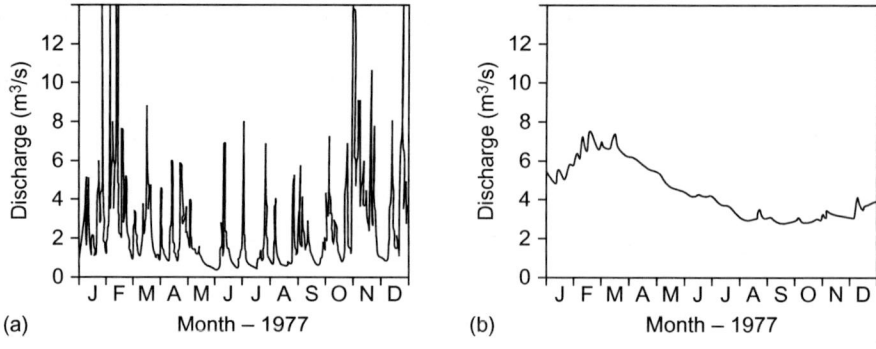

Figure 2.6   Average daily flows of two rivers

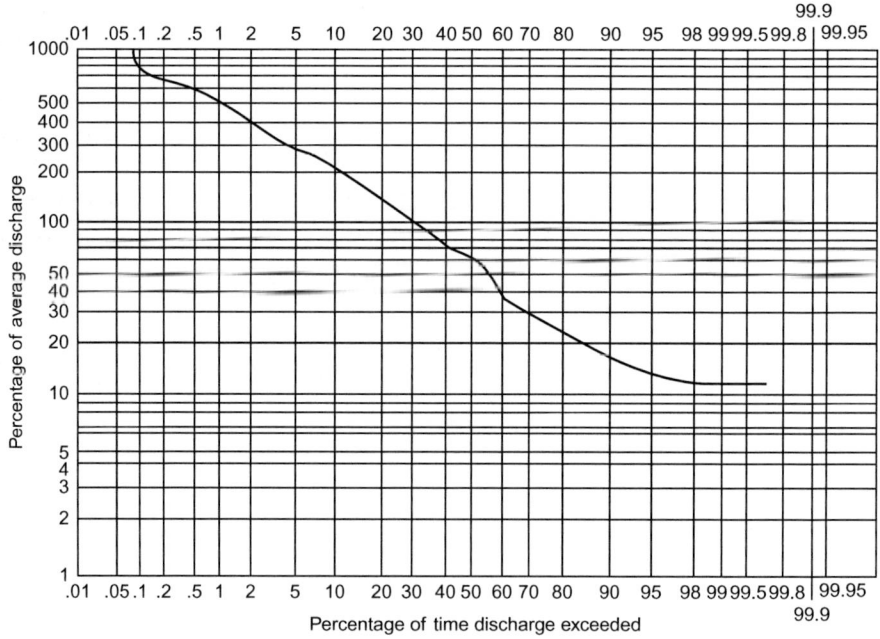

Figure 2.7   Typical flow duration curve

the output of the hydro scheme might correlate with the day-to-day load demand on the power system.

The power output of a hydro turbine is given by the simple expression:

$$P = QH\eta\rho g \tag{2.1}$$

where P = output power (W), Q = flow rate (m^3/s), H = effective head (m), η = overall efficiency, ρ = density of water (1000 kg/m^3) and g = acceleration due to gravity.

A high effective head is desirable as this allows increased power output while limiting the required flow rate and hence cross section of penstock (the pressurised pipe bringing water to the turbine). Various forms of turbine are used for differing combinations of flow rate and head [11]. At lower heads, reaction turbines operate by changing the direction of the flow of water. The water arriving at the turbine runner is still under pressure, and the pressure drop across the turbine accounts for a significant part of the energy extracted. Common designs of reaction turbine include Francis and propeller (Kaplan) types. At higher heads, impulse turbines are used, for example Pelton or Turgo turbines. These operate by extracting the kinetic energy from a jet of water that is at atmospheric pressure. For small hydro units (<100 kW) a cross-flow impulse turbine, where the water strikes the runner as a sheet rather than a jet, may be used. Typically, reasonable efficiencies can be obtained with impulse turbines down to 1/6th of rated flow, whereas for reaction turbines efficiencies are poor below 1/3rd of rated flow. This has obvious implications for rating of plant given the variable resource of some catchments, and these reductions in efficiency with flow rate are taken into consideration when calculating annual energy yield.

Small-scale hydro schemes may use induction of synchronous generators. Low head turbines tend to run more slowly and so either a gearbox or a multi-pole generator is required. One particular design consideration is to ensure that the turbine-generator will not be damaged in the event of the electrical connection to the network being broken and hence the load lost and the turbine-generator over-speeding [8]. This consideration favours the use of robust, squirrel-cage induction generators over wound rotor synchronous machines for simple, small-scale hydro systems.

Increasingly variable speed hydro-generator sets are used to match the operating characteristic of the turbine to the variable flow rates experienced during different hydrological conditions. Just as with variable speed wind turbines, this requires the use of power electronics to interface the generator to the 50/60 Hz network.

2.3.2   Wind power plants

A wind turbine operates by extracting kinetic energy from the wind passing through its rotor. The power developed by a wind turbine is given by:

$$P = \frac{1}{2}C_p\rho V^3 A \tag{2.2}$$

where C_p = power coefficient – a measure of the effectiveness of the aerodynamic rotor, P = power (W), V = wind velocity (m/s), A = swept area of rotor disk (m^2) and ρ = density of air (1.25 kg/m^3).

As the power developed is proportional to the cube of the wind speed, it is important to locate any electricity generating turbines in areas of high mean annual wind speed and the available wind resource is an important factor in determining where wind farms are sited. Often the areas of high wind speed will be away from habitation, and the associated well-developed electrical distribution network, leading to a requirement for careful consideration of the connection of wind turbines to relatively weak electrical distribution networks. The difference in the density of the working fluids (water 1000 kg/m^3 and air 1.25 kg/m^3) shows clearly why a wind turbine rotor of a given rating is so much larger than a hydro turbine. A 2 MW wind turbine will have a rotor diameter of some 60–80 m mounted on a 60–90 m high tower. The force exerted on the rotor is proportional to the square of the wind speed and so the wind turbine must be designed to withstand large forces during high winds. Most modern designs use a three-bladed horizontal-axis rotor as this gives a good value of peak C_p together with an aesthetically pleasing design.

The power coefficient (C_p) is a measure of how much of the energy in the wind is extracted by the turbine rotor. It varies with rotor design and the relative speed of the rotor and the wind (known as the tip speed ratio) to give a maximum practical value of approximately 0.4–0.45[2] [12].

Figure 2.8 is the Power Curve of a wind turbine that indicates its output at various wind speeds. At wind speeds below cut-in (~ 5 m/s), no significant power is developed. The output power then increases rapidly with wind speed until it reaches its rated value and is then limited by some control action of the turbine. This part of the characteristic follows an approximately cubic relationship between wind speed and output power although this is modified by changes in C_p. Then at the shutdown wind speed (25 m/s in this case), the rotor is parked, or allowed to idle at low speed, for safety.

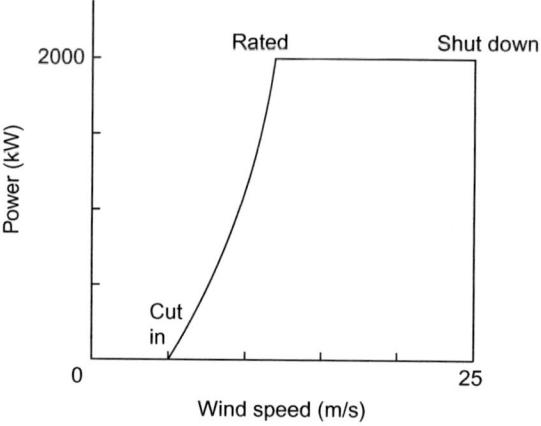

Figure 2.8   Wind turbine Power Curve

Figure 2.9 shows a typical annual distribution of hourly mean wind speeds from a UK lowland site, and by comparison with Figure 2.8, it may be seen that the turbine will only be operating at rated output for some 10–15% of the year. Depending on the site wind speed distribution, the turbine may be shut down due to low winds for up to 25% of the year, and during the remaining period the output will fluctuate with wind speed. Figures 2.8 and 2.9 can be combined to calculate

---

[2] The absolute maximum C_p that any rotor can achieve is known as the Betz Limit and is 16/27 or 0.59.

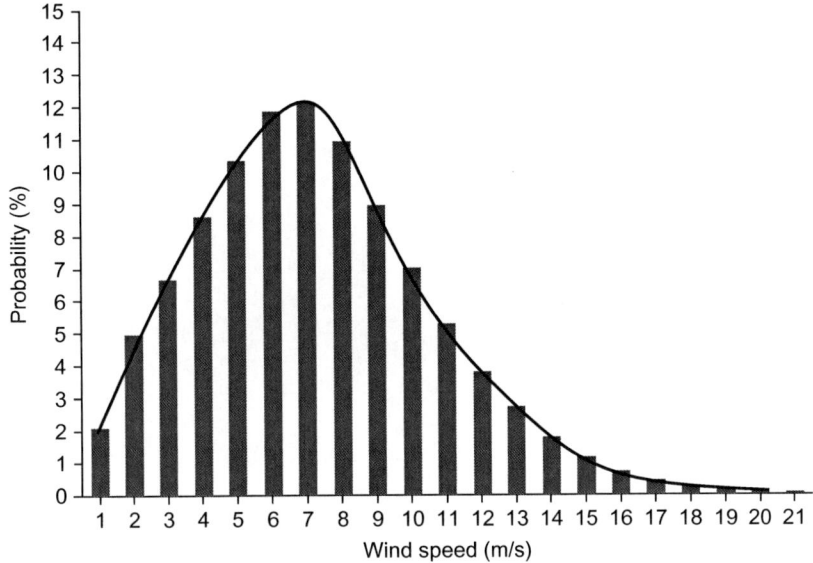

Figure 2.9 Distribution of hourly mean wind speeds of a typical lowland site

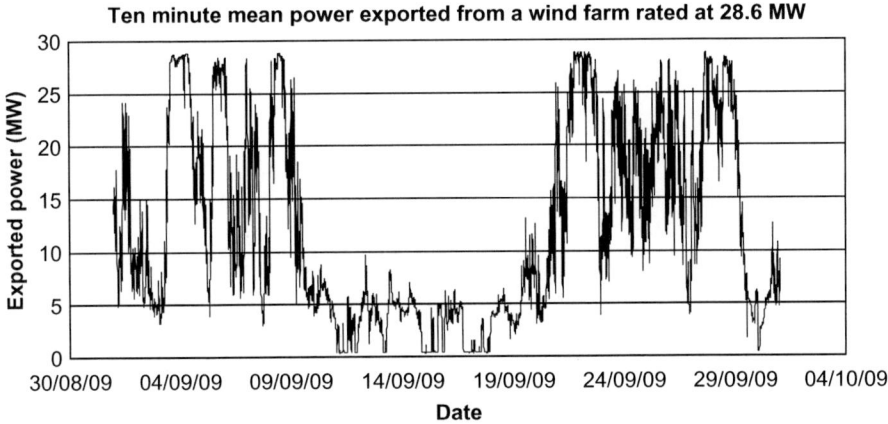

*Figure 2.10 Real power output of a wind farm over one month
[data from RES]*

the annual energy yield from a wind turbine, but they provide no information as to when the energy is generated. Figures 2.10 and 2.11 show time series of the power outputs of a wind farm in the United Kingdom, while Figure 2.12 shows the capacity factors of a number of wind farms by season.

The wind speed distribution of Figure 2.9 is formed from hourly mean wind speeds, and the Power Curve is measured using 10-minute average data. There are also

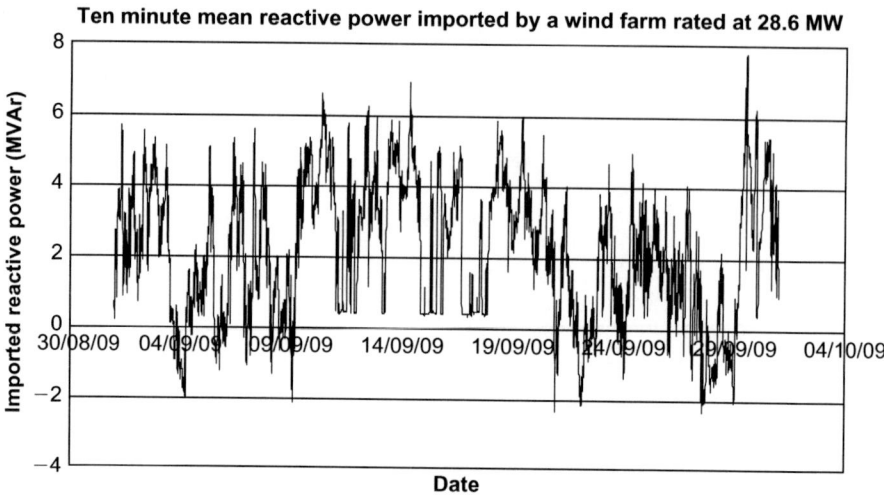

Figure 2.11   Reactive power import of a fixed speed induction generator wind farm over one month [data from RES]

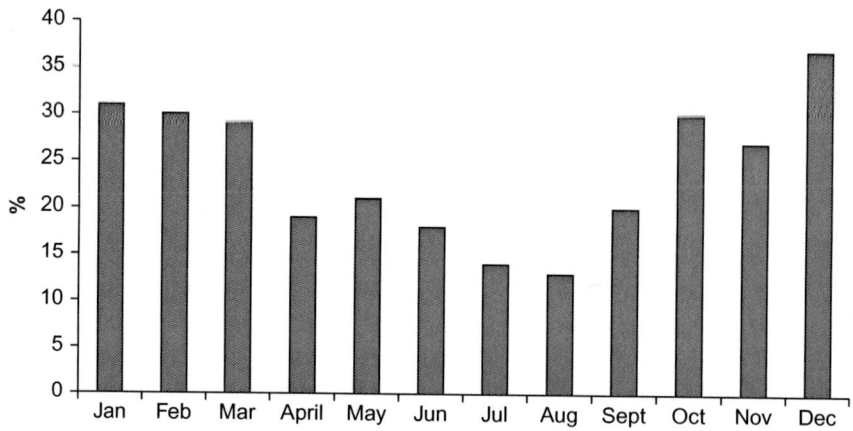

Figure 2.12   Monthly capacity factors of early UK wind farms

important higher frequency effects that, although they do not significantly influence the energy generated, are important for their consequences on the wind turbine machinery and on network power quality. With some designs of fixed speed wind turbines, it has been found to be rather difficult to limit the output power to the rated value indicated by the horizontal section of the Power Curve (Figure 2.8), and transient overpowers of up to twice nominal output have been recorded in some instances. Therefore, when considering the impact of the turbines on to the distribution network, it is important to establish what the maximum transient output power is likely to be.

The torque from a horizontal-axis wind turbine rotor contains a periodic component at the frequency at which the blades pass the tower. This cyclic torque is due to the variation in wind speed seen by the blade as it rotates. The variation in wind speed is due to a combination of: tower shadow, wind shear and turbulence. In a fixed speed wind turbine, this rotor torque variation is then translated into a change in the output power and hence a voltage variation on the network. These dynamic voltage variations are often referred to as 'flicker' because of the effect they have on incandescent lights. The human eye is very sensitive to changes in light intensity particularly if the variation occurs at frequencies around 10 Hz. For a large wind turbine the blade passing frequency will be around 1–2 Hz, and although the eye is less sensitive at this frequency, it will still detect voltage variations greater than about 0.5%. In general the torque fluctuations of individual wind turbines in a wind farm are not synchronised and so the effect in large wind farms is reduced as the variations average out.

Although reliable commercial wind turbines can be bought from a variety of manufacturers, there is still considerable development of the technology particularly as the size and ratings of turbines increase. Some of the major differences in design philosophy include: (1) fixed or variable speed operation, (2) direct drive generators or the use of a gearbox and (3) stall or pitch regulation [12,13].

Fixed speed wind turbines using induction generators are simple, and it may be argued, robust. It is not usual to use synchronous generators on network-connected fixed speed wind turbines, as it is not practicable to include adequate damping in a synchronous generator rotor to control the periodic torque fluctuations of the aerodynamic rotor. Some very early wind turbine designs did use synchronous generators by including mechanical damping in the drive trains (e.g. by using a fluid coupling), but this is no longer a common practice. Figure 2.13(a) shows a simplified schematic representation of a fixed speed wind turbine. The aerodynamic rotor is coupled to the induction generator via a speed increasing gearbox. The induction generator is typically wound for 690 V, 1000 or 1500 rpm operation. Pendant cables within the tower connect the generator to switched power factor correction capacitors and an anti-parallel soft-start unit located in the tower base. It is common to by pass the soft-start thyristors once the generator flux has built up, excited from the network voltage. The power factor correction capacitors are either all applied as soon as the generator is connected or they are switched in progressively as the average output power of the wind turbine increases. A local transformer, typically 690 V/33 kV in UK wind farms is located either inside the tower or adjacent to it.

With variable speed operation it is possible, in principle, to increase the energy captured by the aerodynamic rotor by maintaining the optimum power coefficient over a wide range of wind speeds, and perhaps more importantly, also reduce mechanical loads. However, it is then necessary to decouple the speed of the rotor from the frequency of the network through some form of power electronic converter. A simple approach (Figure 2.13(b)) is to use a wound rotor induction generator with controllable, external rotor resistors. The external resistors alter the shape of the torque–speed curve and so allow the generator, and rotor, speed to increase by up to 10%. An obvious disadvantage is that power is dissipated in the resistors as heat and so wasted. Hence a development was to use a wound rotor induction generator but with back/back voltage source converters in the rotor circuit (Figure 2.13(c), DFIG, doubly fed induction generator). Power flows from the generator out to the network when the generator operates above synchronous speed but back towards the rotor at below synchronous speed [13,14].

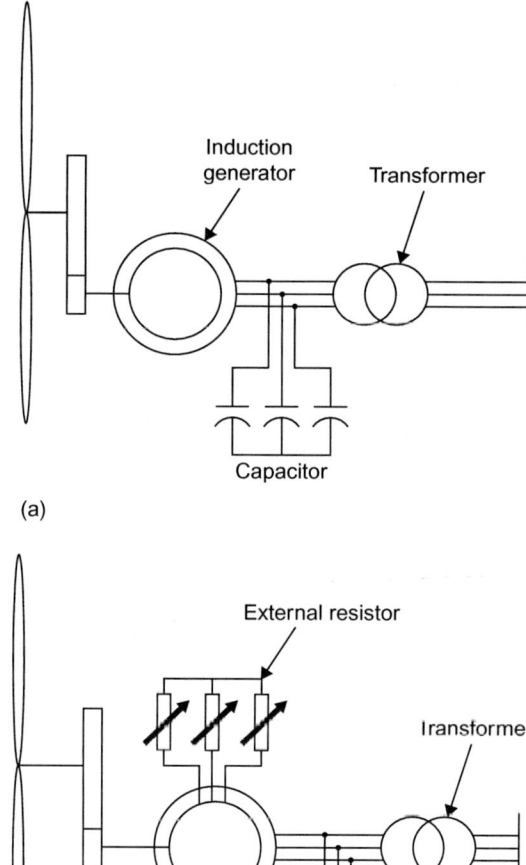

Figure 2.13   *Wind turbine architectures [14]. (a) Fixed speed induction generator wind turbine. (b) Variable slip wind turbine. (c) Doubly fed induction generator wind turbine. (d) Full power converter wind turbine. (e) Full power converter wind turbine (diode rectifier).*

If a wide range of variable speed operation is required, then the arrangement shown in Figure 2.13(d) may be used. Two voltage source converter bridges are used to interface the wind turbine drive train to the network. The network-side bridge is commonly used to maintain the voltage of the DC link and control reactive power flows with the network. The generator-side converter is used to control the torque on the rotor and

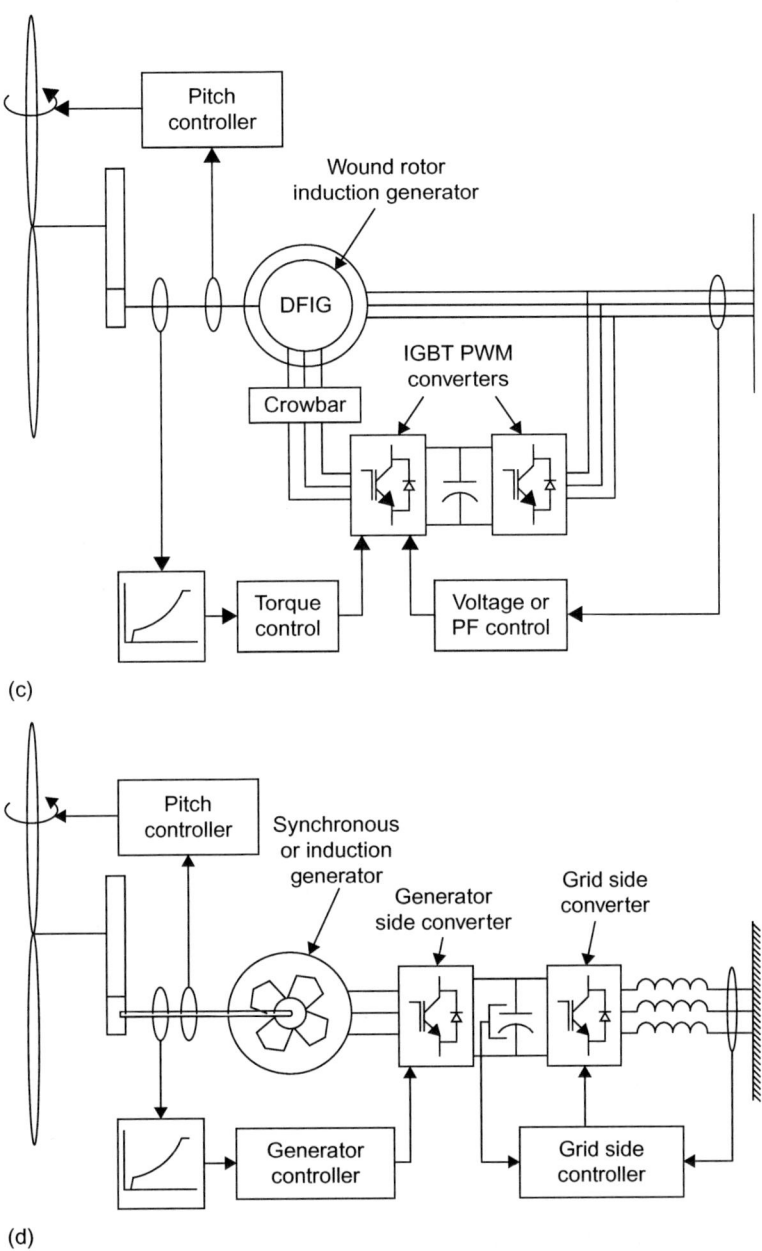

(c)

(d)

Figure 2.13   Continued

hence output power of the wind turbine. The generator may be either synchronous or induction, and some form of pulse width modulated (PWM) switching pattern is used on the converter bridges. As all the power is transferred through the variable speed equipment, both the converters are rated at the full power of the generator. Hence there are

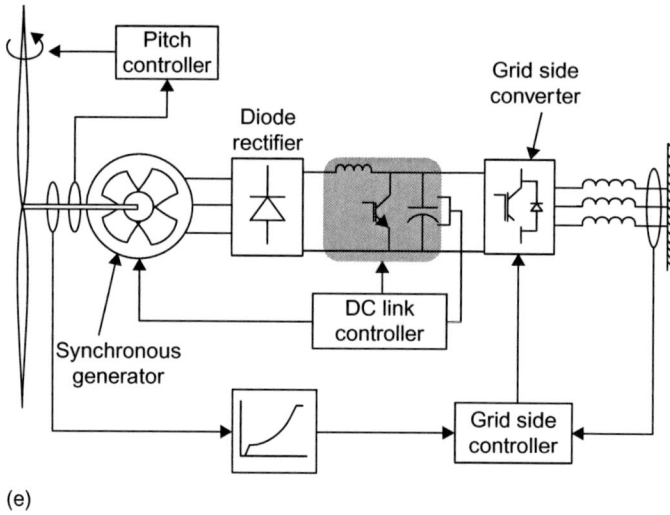

(e)

Figure 2.13   Continued

significant electrical losses, and at low output powers, the energy gains from variable speed operation of the rotor may not be fully realised. However, variable speed operation gives a number of other important advantages that follow from the variable speed rotor acting as a flywheel type energy store. These include: (1) a reduction in mechanical loads that allows lighter mechanical design and (2) smoother output power. Some designs operate over a restricted speed range (e.g. the DFIG typically has a speed range of ±30%) in order to gain most of the benefits of variable speed operation but with lower electrical losses in the smaller converters. The power converters of variable speed wind turbines allow independent control of real and reactive power and so facilitate compliance with the requirements of the network operators.

Some manufacturers have dispensed with the gearbox between the rotor and the generator and have developed large diameter direct drive generators that rotate at the same speed as the aerodynamic rotor. It is not practical to use large diameter induction generators, the air gap required is too small to be constructed over such a wide diameter, and so either wound rotor or permanent magnet synchronous generators are used. These multi-pole generators are then interfaced to the network as shown in Figure 2.13(d) or (e).

Once rated wind speed is reached it is necessary to limit the power into the wind turbine rotor and so some form of rotor regulation is required. Stall regulation is a passive system with no moving parts. The rotor blades enter aerodynamic stall once the wind speed exceeds the rated value. Stall regulation relies on the rotational speed of the rotor being controlled and so is usually found on fixed speed wind turbines. Pitch-regulated rotors have an actuator and control system to rotate the blades about their axes and so limit power by reducing the angle of attack seen by the aerofoil. Pitch regulation requires a more complex control system but can lead to rather higher energy capture. A combination of these two approaches is known as active-stall, where the blade is rotated about its axis, but the main power limitation remains aerodynamic stall.

These differences in design philosophy do not fundamentally affect the basic energy production function of a wind turbine, which is defined by Figures 2.8 and 2.9, but they do have an important influence on the dynamic operation of the turbine and the output power quality. They also determine whether the wind turbine can ride through network faults (fault ride through), which is an important requirement of many transmission system operators [15].

2.3.3   Offshore wind energy

Due to difficulties in obtaining planning permission (permitting) for wind turbines onshore, there is major interest in siting them offshore. A number of offshore wind farms of rating 50–100 MW have been commissioned with plans for installations of up to 1000 MW now well advanced.

The advantages of offshore installations include:

- reduced visual impact
- higher mean wind speed
- reduced wind turbulence
- low wind shear leading to lower towers

The disadvantages of offshore installations include:

- higher capital costs
- access restrictions in poor weather
- requirement of submarine cables and offshore transformer stations for the larger wind farms are required

The turbines used offshore to date have been similar to those used on land but with increased physical protection against the harsh marine environment. However, as the production volumes increase, it is likely that the turbine designs will become tailored for offshore with greater emphasis on reliability and also perhaps increased rotor tip speeds. Tip speeds are limited on terrestrial wind turbines as very high tip speeds lead to excessive noise.

The generating voltage within the wind turbines is typically 690 V with the wind farm power collection circuits at around 30 kV. Generally the electrical connection to shore has used alternating current (AC) at around 30 kV for the smaller offshore wind farms increased to 150 kV using an offshore transformer station for the larger installations located further offshore (Figure 2.14). For large wind farms located a long way offshore, it is likely that the voltage source high-voltage direct current (HVDC) transmission will be used (Figure 2.15). The capacitance of the long submarine cables used in large AC-connected wind farms can generate excessive reactive power and hence lead to voltage rise. The very long cable lengths can also give difficulties with unusual harmonic resonances and over-voltages caused by switching transients.

2.3.4   Solar photovoltaic generation

Photovoltaic generation, or the direct conversion of sunlight to electricity, is a well-established technology with a number of major manufacturers producing

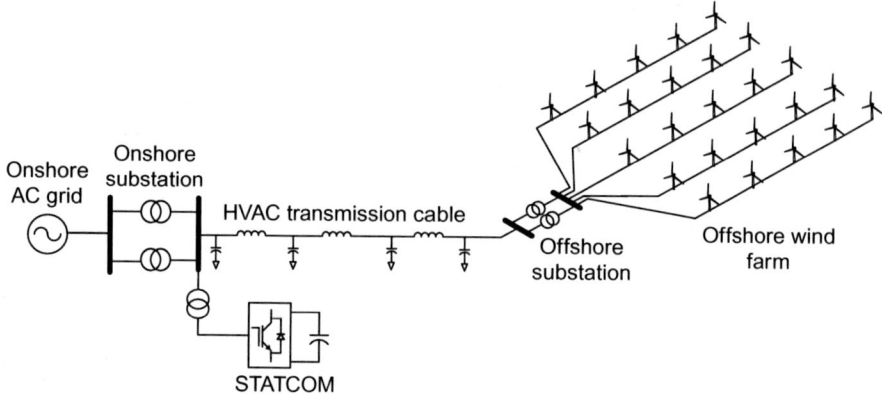

Figure 2.14   *Offshore wind farm using AC transmission to shore (note the STATCOM to control reactive power and hence voltage of the onshore network)*

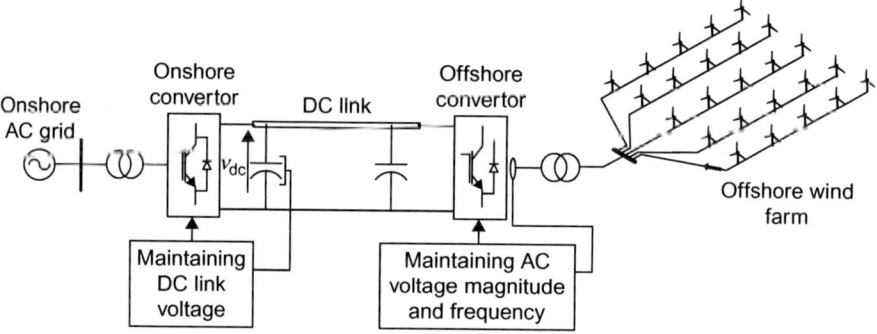

Figure 2.15   *Offshore wind farms using voltage source HVDC transmission to shore*

equipment. For many years, its main application was off-grid for high-value, small electrical loads that were a long way from the nearest distribution network (e.g. vaccine refrigerators and remote communication systems). More recently, stimulated by support through feed-in-tariffs its use as grid-connected distributed generation has increased dramatically particularly in Germany and Spain, while Japan also has a very active programme of installation.

A number of large (megawatt scale) demonstration projects have been constructed in the past, but interest in Europe is now focused on smaller installations. The photovoltaic modules may be roof mounted or incorporated into the fabric of buildings in order to reduce overall cost and space requirements. Thus, these rather small PV installations (typically from 1 to 50 kW) are connected directly at customers' premises and so to the LV distribution network. This form of generation is truly distributed with very large number of residences and commercial buildings being equipped with photovoltaic generators.

Outside the earth's atmosphere, the power density of the solar radiation, on a plane perpendicular to the direction to the sun, is approximately 1350 W/m². As the sunlight passes through the earth's atmosphere the energy in certain wavelengths is selectively absorbed. Thus both the spectrum and power density change as the light passes through the atmosphere.

A term known as the air mass (AM) has been introduced to describe the ratio of the path length of the solar radiation through the atmosphere to its minimum value, which occurs when the sun is directly overhead. For the standard testing of photovoltaic modules, it is conventional to use a spectrum corresponding to an air mass of 1.5 (AM 1.5) but with the power density adjusted to 1000 W/m². A cell temperature of 25 °C is also assumed. These are the conditions under which the output of a photovoltaic module is specified and tested in the factory. They may differ significantly from those experienced in service.

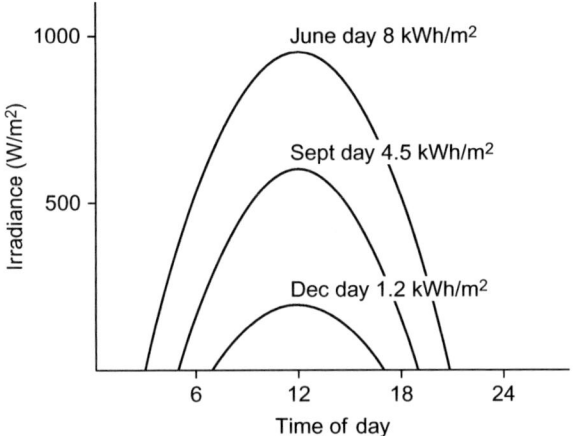

Figure 2.16   Solar irradiance (and daily solar energy) on a flat plane at 48° north

The total or global solar energy arriving at a photovoltaic module is made up of direct and diffuse components. Direct radiation consists of the almost parallel rays that come from the solar disc. Diffuse radiation is that scattered in the atmosphere and approaches the module from all parts of the sky. On a clear day the direct component may make up 80–90% of the total radiation, but on a completely overcast day this drops to almost zero leaving a small diffuse component, say 10–20%, of the radiation expected on a clear day. Figure 2.16 shows typical daily curves of global irradiance on a horizontal surface for clear conditions in latitude 48° and illustrates the difference in radiation (and daily solar energy) in the winter and summer months. Mainland England and Scotland extends from latitude 50° to 59° and so the seasonal variations in the solar resource will be even more

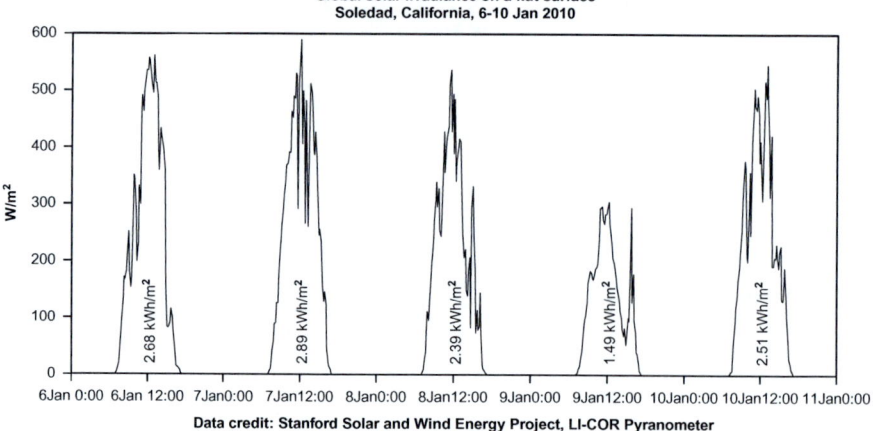

Figure 2.17   *Global irradiance on a flat surface in California over 5 days*

pronounced in the United Kingdom. Figure 2.17 shows the effect of cloud on solar irradiance in California.

A description of the physics of photovoltaic energy conversion is given by a number of authors, for example Green [16], Van Overstraeton and Mertens [17] and Markvart [18]. However, an initial understanding of the performance of a solar cell may be obtained by considering it as a diode in which light energy, in the form of photons with the appropriate energy level, falls on the cell and generates electron-hole pairs. The electrons and holes are separated by the electric field established at the junction of the diode and are then driven round an external circuit by this junction potential. There are losses associated with the series and shunt resistance of the cell as well as leakage of some of the current back across the p–n junction. This leads to the equivalent circuit of Figure 2.18 and the operating characteristic of Figure 2.19. Note that Figure 2.19 is drawn to show the comparison with a conventional diode characteristic. The current produced by a solar cell is proportional to its surface area and the incident irradiance, while the voltage is limited by the forward potential drop across the p–n junction.

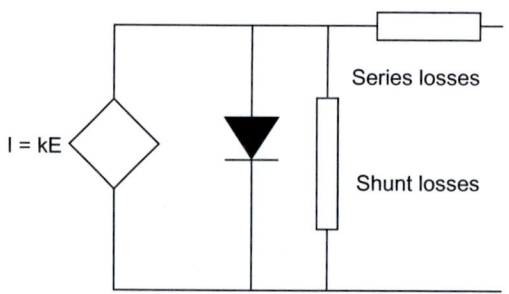

Current source output is proportional to solar irradiance, i.e. $I = kE$.

Figure 2.18   *Equivalent circuit of a PV cell*

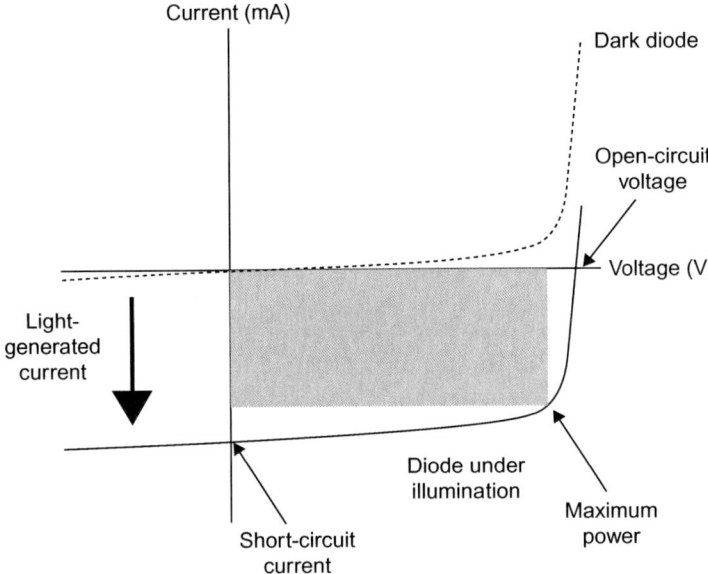

Figure 2.19   Voltage/Current characteristic of an illuminated silicon diode

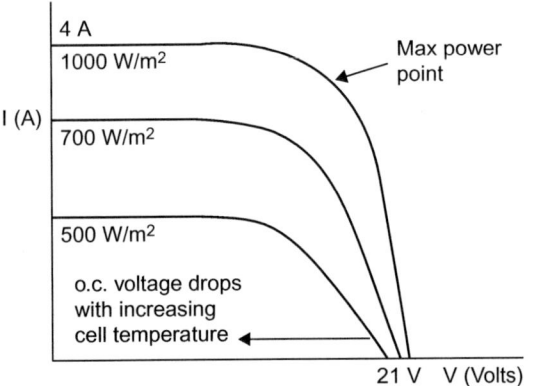

Figure 2.20   Typical voltage/current characteristic of a PV module

In order to produce higher voltages and currents the cells are arranged in series/parallel strings and packaged into modules for mechanical protection. These robust and maintenance-free modules then have an electrical characteristic by convention, as shown in Figure 2.20.

The maximum power output of the module is obtained near the knee of its voltage/current characteristic. However, the output voltage, but not the current, of the module is reduced at increased cell temperature and so the maximum output power can only be obtained using a maximum power point tracking (MPPT) stage on the input to the converter. A number of techniques are used to maintain operation at the maximum power point of the characteristic including Hill Climbing, known sometimes as Perturb and Observe, whereby the inverter makes a small change (e.g. increase) in the voltage applied to the module and looks to see if more power is obtained. If more power is obtained from the module, the voltage is increased further, if not a small decrease in voltage is made. This operates continuously.

Figure 2.21 shows the schematic diagram of an inverter for a small PV 'grid-connected' system. (Note that the term 'grid-connected' is often used rather loosely to describe small distributed generation systems, which are connected to a local utility distribution network and not to the interconnected high voltage interconnected grid network.) The inverter typically consists of: (1) an MPPT circuit, (2) an energy storage element, usually a capacitor, (3) a DC:DC converter to increase the voltage, (4) a DC:AC inverter stage, (5) an isolation transformer to ensure DC is not injected into the network and (6) an output filter to restrict the harmonic currents passed into the network, particularly those near the device switching frequencies. Very small inverters (up to say 200 W) may be fitted to the back of individual modules, the so-called 'AC module' concept or larger inverters used for a number of modules, the 'string inverter' concept. Usually PV inverters operate at unity power factor (producing only real power, W and not reactive power, VARs). They do not take part in system voltage control with the low X/R ratio of LV distribution circuits leading to reactive power flows having a very limited effect on the magnitude of network voltage.

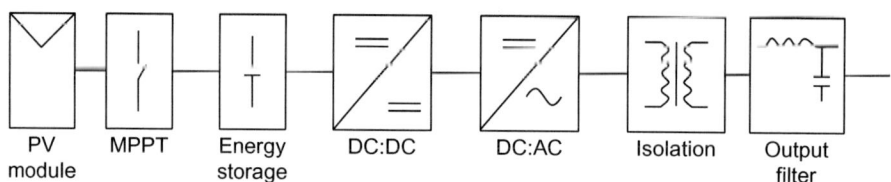

| PV module | MPPT | Energy storage | DC:DC | DC:AC | Isolation | Output filter |

Figure 2.21   Schematic representation of a small PV inverter for 'grid-connected' operation

Although all current commercial photovoltaic cells operate on the same general principles, there are a number of different materials used. The early cells used mono-crystalline silicon, and this is still in common use. Very large single crystals of silicon are grown as cylinders and then cut into circular wafers and doped. The single crystal is expensive to form but allows high efficiencies, and overall module efficiencies up to 20% may be obtained. An alternative technique, again using bulk silicon, is to cast poly-crystalline cubes and then cut these into square wafers. Although a cheaper process, poly-crystalline modules are typically some 4% less efficient due to the random crystal structure of the cells. Both mono- and poly-crystalline silicon cells are in general use and the choice between them is generally made on commercial grounds.

The bulk, purified silicon is expensive and so much effort has been expended on the 'thin film' devices where only a small volume of active material is deposited on a cheaper, inert substrate. The active materials commonly used include amorphous silicon and cadmium telluride, but a large number of other materials have been investigated. Thin film materials are less efficient than bulk silicon, and in the past, the performance of some cells degraded over time. However, thin film solar cells are extensively used in consumer products and are offered by a number of manufacturers for general application. Research is under way on the 3rd generation of solar cells using rather different principles, but these are not yet commercially significant.

2.4 Summary

There is an increasing number of types of generating plant being connected to distribution networks. This distributed generation offers considerable environmental and economic benefits including high overall efficiencies, the use of renewable energy sources and reduction in greenhouse gas emissions. The location and rating of the generators is determined either by the heat loads or by the renewable energy resource. Renewable energy generation is operated in response to the available resource in order to obtain the maximum return on the, usually high, capital investment. CHP plant is usually operated in response to the needs of the host facility or a district-heating load. Although it can be operated with regard to the needs of the power system, this is likely to reduce its efficiency unless, as in the case of the Danish rural CHP schemes, some form of energy storage is used.

It is, of course, possible to install distributed generation, for example small gas turbines, reciprocating internal combustion engines or even battery storage systems, specifically in order to support the power system. This was done in France with some 610 MW of installed diesel engine capacity distributed throughout the distribution network. This can then be centrally dispatched to contribute to either local or national deficits of power.

As more and more distributed generation is operated on the power system, it will displace increasingly the large central generation that at present provides the control of the system. Then distributed generators will need to be operated flexibly to ensure the load and supply of the power system is balanced and the ancillary services are provided.

References

1. UK Government Statistical Office. *Digest of UK Energy Statistics 2009*. Published by the Stationary Office. Available from http://www.decc.gov.uk/en/content/cms/statistics/publications/dukes/dukes.aspx [Accessed March 2010].
2. Horlock J.H. *Cogeneration: Combined Heat and Power Thermodynamics and Economics*. Oxford: Pergamon Press; 1987.
3. Jorgensen P., Gruelund Sorensen A., Falck Christensen J., Herager P. Dispersed CHP Units in the Danish Power System. Paper No. 300-11. Presented at CIGRE Symposium 'Impact of Demand Side Management, Integrated

Resource Planning and Distributed Generation;' Neptun, Romania, 17–19 Sep. 1997.

4. Marecki J. *Combined Heat and Power Generating Systems*. London: Peter Peregrinus; 1988.

5. Khartchenko N.V. *Advanced Energy Systems*. Washington: Taylor and Francis; 1998.

6. Packer J., Woodworth M. 'Advanced package CHP unit for small-scale generation'. *IEE Power Engineering Journal*. May 1991, pp. 135–142.

7. Hu S.D. *Cogeneration*. Reston, VA: Reston Publishing Company; 1985.

8. Allan C.L.C. Water-Turbine-Driven Induction Generators. IEE Paper No. 3140S, Dec. 1959.

9. Tong J. *Mini-Hydropower*. Chichester: John Wiley and Sons; 1997.

10. Fraenkel P., Paish O., Bokalders V., Harvey A., Brown A., Edwards R. *Micro-Hydro Power, a Guide for Development Workers*. London: IT Publications; 1991.

11. Boyle G. (ed.). *Renewable Energy*. 2nd edn. Oxford: Oxford University Press; 2004.

12. Manwell J.G., McGowan F., Rogers A.L. *Wind Energy Explained*. Chichester: John Wiley and Sons; 2002.

13. Heier S. *Grid Integration of Wind Energy Conversion Systems*. 2nd edn. Chichester: John Wiley and Sons; 2006.

14. Anaya-Lara O., Jenkins N., Ekanayake J., Cartwright P., Hughes M. *Wind Energy Generation; Modelling and Control*. Chichester: John Wiley and Sons; 2009.

15. Freris L.L., Infield D. *Renewable Energy in Electric Power Systems*. Chichester: John Wiley and Sons; 2008.

16. Green M.A. *Solar Cells*. Englewood Cliffs, NJ: Prentice Hall; 1982.

17. Van Overstraeton R.J., Mertens R.P. *Physics, Technology and Use of Photovoltaics*. Bristol, UK: Adam Hilger; 1986.

18. Markvart T. (ed.). *Solar Electricity*. Chichester: John Wiley and Sons; 2000.

Chapter 3
Distributed generators and their connection to the system

Solar Array, RES Group Head Office – Beaufort Court, Hertfordshire, England Hybrid thermal/photovoltaic solar energy panels. The system produces electrical energy and hot water [RES]

3.1 Introduction

The connection of distributed generators to the network requires understanding the operation and control of different types of generating plant, and often needs studies to evaluate the performance of the power system with the new generation, under both normal and abnormal operating conditions. The types of generators used for distributed generation depend on their application and energy source. For example a small diesel generator set would normally use a synchronous generator while a wind turbine may employ a squirrel-cage induction generator, called a fixed speed induction generator (FSIG) in this book, a doubly fed induction generator (DFIG) or full power converter (FPC) connected generator. DC sources or those generating at high frequency such as photovoltaic systems, fuel cells or micro-turbines require a power electronic converter to interface them into the power system. The performance and characteristics of these different types of power plant differ significantly.

The performance of networks with distributed generation is studied using computer programs:

- load or power-flow programs to evaluate the steady-state voltages at busbars and power flows in circuits;
- fault calculators to determine the fault currents caused by different types of faults; and
- stability studies to determine the stability of the power system and/or distributed generators, normally following a fault.

3.2   Distributed generators

3.2.1   *Synchronous generators*

Synchronous generators are always used in large conventional power plants. The principles of operation of synchronous generators are described in detail in a large number of excellent textbooks, for example References 1–5. Large steam turbine generator sets use turbo-alternators consisting of a cylindrical rotor with a single DC winding to give one pair of poles and hence maximum rotational speed (3000 rpm on a 50 Hz system; 3600 rpm on a 60 Hz system). Hydro generators usually operate at lower speeds, and then use multiple pole generators with a salient-pole rotor. Smaller engine-driven units also generally use salient-pole generators.

Permanent magnet generators are not commonly found in large generating units, as although higher efficiencies can be achieved, direct control of the rotor magnetic field is not possible. However, permanent magnet synchronous generators are used in some variable-speed wind turbines and in some microgenerators. Also, permanent magnet generators can be designed to withstand the mechanical forces of high-speed operation when driven by micro-turbines.

For power system studies, synchronous generators are represented by a voltage behind an impedance as shown in Figure 3.1. In this model X_s is the synchronous reactance and R is the stator resistance. The synchronous impedance is then, $Z_s = R + jX_s$. For large generators, R is often neglected.

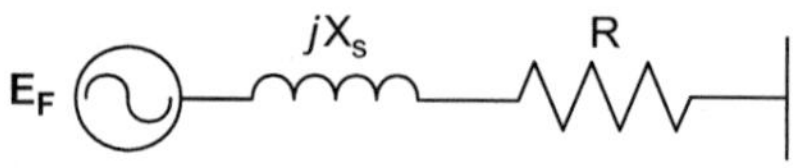

Figure 3.1   Voltage behind impedance model of a synchronous generator

Figure 3.2 shows a simplified representation of a 5 MW synchronous distributed generator driven by a small steam turbine. If the short-circuit level at the point of connection (C) is, say, 100 MVA, with an X/R ratio of, say, 10 then the total source impedance on a 100 MVA base will be approximately

$$Z = 0.1 + j1.0 \, \text{pu}$$

and with a realistic value of X_s of 1.5 pu on the machine rating then, again on 100 MVA base

$$X_s = j30 \, \text{pu}$$

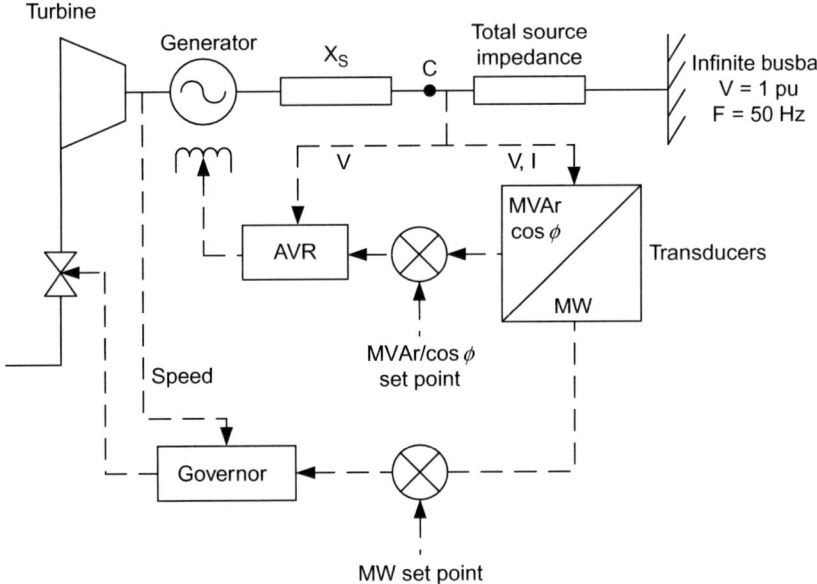

Figure 3.2   Control of a synchronous distributed generator

Thus it can be seen immediately that $|X_s| \gg |Z|$ and to a first approximation, the synchronous generator will have a very small effect on network voltage (point C). As a small generator cannot affect the frequency of a large interconnected power system, then the distributed generator can be considered to be connected directly to an infinite busbar.

Figure 3.2 is an over-simplification in one important respect in that the other loads on the network are not shown explicitly and these may alter the voltage at the point of connection of the generator considerably. In some smaller power systems, changes in total system load or outages on the bulk generation system will also cause significant changes in frequency.

A conventional method of controlling the output power of a generating unit is to set up the governor on a frequency/power droop characteristic. This is shown in Figure 3.3 where the line (a–b) shows the variation in frequency (typically 4%) required to change the power output of the prime mover from no-load to full-load. Thus with 1 per unit (pu) frequency (50 Hz) the set will produce power P1. If the frequency falls by 1% the output power increases to P3 while if the system frequency rises by 1% the output power reduces to P2. This, of course, is precisely the behaviour required from a large generator that can influence the system frequency; if the frequency drops more power is required while if the frequency rises less power is needed. The position of the droop line can be changed vertically along the y-axis and so by moving the characteristic to (a′–b′) the power output can be restored to P1 even with an increased system frequency or by moving to (a″–b″) for a reduced system frequency.

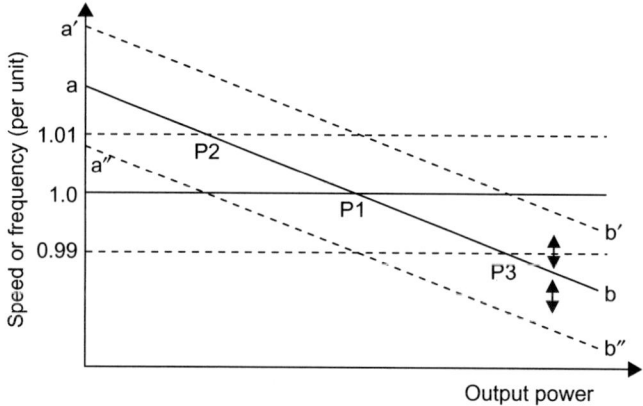

Figure 3.3   Conventional governor droop characteristic for generator governor control

A similar characteristic can be set up for voltage control (Figure 3.4) with the axes replaced by reactive power and voltage. Again, consider the droop line (a–b). At 1 pu voltage no reactive power is exchanged with the system (operating point Q1). If the network voltage rises by 1% then the operating point moves to Q2 and reactive power is imported by the generator, in an attempt to control the voltage rise. Similarly if the network voltage drops the operating point moves to Q3 and reactive power is exported to the system. Translating the droop lines to (a′–b′) or (a″–b″) allows the control to be reset for different condition of the network. The slope of both the frequency and voltage droop characteristics can also be changed if required. These frequency and voltage droop characteristics describe simple proportional control systems. In practice, governor and automatic voltage regulator (AVR) controls are much more complex and will include integral terms to eliminate steady-state error. However, the principle remains that this type of controller is intended to control the network variables (i.e. frequency or voltage) and so is appropriate for larger generators.

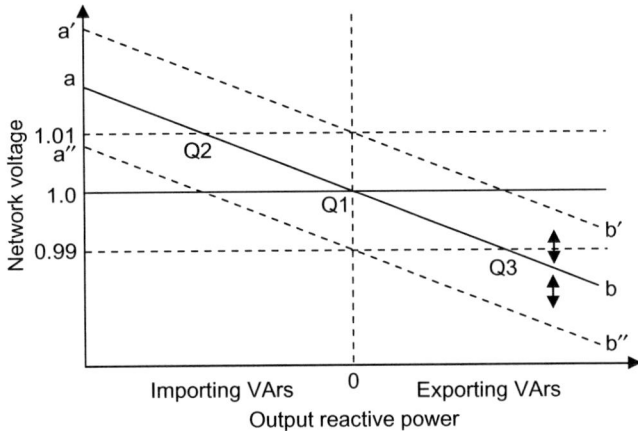

Figure 3.4 Quadrature droop characteristic for generator excitation control

These types of control schemes may not be appropriate for small-distributed synchronous generators. For example an industrial combined heat and power (CHP) plant may wish to operate at a fixed power output, or fixed power exchange with the network, irrespective of system frequency. Similarly, operation with no reactive power exchange with the network may be desirable in order to minimise reactive power charges. If the generators are operated on the simple droop characteristics, illustrated in Figures 3.3 and 3.4, then both real and reactive power outputs of the generator will change constantly as the network frequency and voltage varies due to external influences.

For relatively small synchronous distributed generators on strong networks, control is often based on real and reactive power output rather than on frequency and voltage as might be expected in stand-alone installations or for large generators. As shown in Figure 3.2, voltage and current signals are obtained at the terminals of the generator and passed to transducers to measure the generated real and reactive power output. The main control variables are MW, for real power, and MVAr or $\cos \phi$ for reactive power. A voltage measurement is also supplied to the AVR and a speed/frequency measurement to the governor, but these are supplementary signals only. It may be found convenient to use the MW and MVAr/$\cos \phi$ error signals indirectly to translate the droop lines and so maintain some of the benefits of the droop characteristic, at least during network disturbances, but this depends on the internal structure of the AVR and governor.

However, the principal method of control is that, for real power control, the measured (MW) value is compared to a set point and then the error signal fed to the governor which, in turn, controls the steam supply to the turbine. In a similar manner the generator excitation is controlled to either an MVAr or $\cos \phi$ setting. The measured variable is compared to a set point and the error passed to the AVR and exciter. The exciter then controls the field current and hence the reactive power output.

It should be noted that the control scheme shown in Figure 3.2 pays no attention to the conditions on the power system. The generator real power output is controlled to a set point irrespective of the frequency of the system, while the reactive power is controlled to a particular MVAr value or power factor irrespective of network voltage. Clearly for relatively large distributed generators, or groups of smaller distributed generators, which can have an impact on the network this is unsatisfactory and more conventional control schemes that provide voltage support are likely to be appropriate [6]. These are well-established techniques used wherever a generator has a significant impact on the power system but there remains the issue of how to influence the owners/operators of distributed generation plant to apply them. Operating at non-unity power factor increases the electrical losses in the generator while varying real power output in response to network frequency will have implications for the prime mover and steam supply, if it is operated as a CHP plant. As increasing numbers of small distributed generators are connected to the network it will become important to coordinate their response both to steady-state network conditions and during disturbances. This requirement for the distributed generation to provide network support is already evident in the transmission network connection Grid Codes that are applied to the connection of large wind farms. These require that large wind farms operate under voltage control (rather than reactive power or power factor control) to maintain the local voltage particularly during network disturbances and also have the capability of contributing to system frequency response. It is likely that as distributed generation becomes an ever more significant fraction of the generation on the power system, then such requirements will become more widely applied.

3.2.2   Induction generators

An induction generator is, in principle, an induction motor with torque applied to the shaft, although there may be some modifications made to the electrical machine design to optimise its performance as a generator. Hence, it consists of an armature winding on the stator, and generally, a squirrel-cage rotor. Squirrel-cage induction machines are found in a variety of types of small generating plant and are always used in fixed speed wind turbines. Wound rotor induction machines are used in some specialised distributed generating units particularly with variable slip, where the rotor resistance is varied by an external circuit, and doubly fed variable-speed wind turbines, where the energy flow in or out of the rotor circuit is controlled by power electronics.

The main reason for the use of squirrel-cage induction generators in fixed speed wind turbines is the damping they provide for the drive train (see Figure 3.5) although additional benefits include the simplicity and robustness of their construction and the lack of requirement for synchronising. The damping is provided by the difference in speed between rotor and the stator magneto motive force (mmf) (the slip speed), but as induction generators increase in size their natural slip decreases [7] and so the transient behaviour of large induction generators starts to resemble that of synchronous machines. Induction generators have also been used in small hydro sets for many years. Reference 8 describes very clearly both the

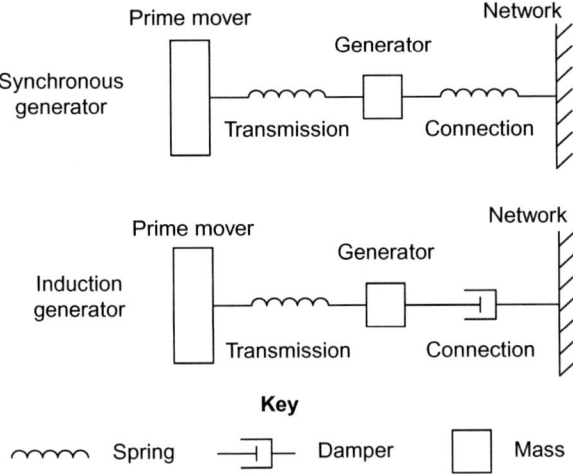

Figure 3.5   Simple mechanical analogues of generators

basic theory of induction generators and their application in small hydro generators in Scotland in the 1950s.

In order to improve the power factor, it is common to fit local power factor correction (PFC) capacitors at the terminals of the generator (Figure 3.6). These have the effect of shifting the circle diagram, as seen by the network, downwards along the y-axis. It is conventional to compensate for all or part of the no-load reactive power demand although, as real power is exported, there is additional reactive power drawn from the network.

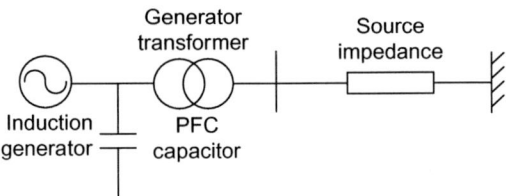

Figure 3.6   Induction generator connected to the infinite bus

An isolated induction generator cannot produce a terminal voltage, as there is no source of reactive power to develop the magnetic field. Hence, when an induction generator is connected to the network there is an initial magnetising inrush transient, similar to that when a transformer is energised, followed by a transfer of real (and reactive) power to bring the generator to its operating speed. For a large distributed induction generator the voltage transient caused by direct-online starting is likely to be unacceptable. Therefore, in order to control both the magnetising inrush and subsequent transient power flows to accelerate or decelerate the generator and prime mover it is common to use a 'soft-start' circuit (Figure 3.7). This merely consists of

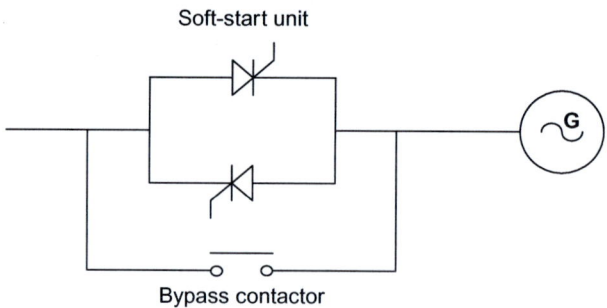

Figure 3.7   Soft-start unit for an induction generator (one phase only shown) [9]

a back-to-back pair of thyristors that are placed in each phase of the generator connection. The soft start is operated by controlling the firing angle of the thyristors so building up the flux in the generator slowly and then also limiting the current that is required to accelerate the drive train. Once the full voltage has been applied, usually over a period of some seconds, the bypass contactor is closed to eliminate any losses in the thyristors. These soft-start units can be used to connect either stationary or rotating induction generators, and with good control circuits, can limit the magnitudes of the connection currents to only slightly more than full-load current. Similar units are, of course, widely used for starting large induction motors.

If a large induction generator, or a number of smaller induction generators, is connected to a network with a low short-circuit level then the source impedance, including the effect of any generator transformers, can become significant. Hence, the equivalent circuit (of Tutorial Chapter II) can be extended, as shown in Figure 3.8 to include the source impedance in the stator circuit.

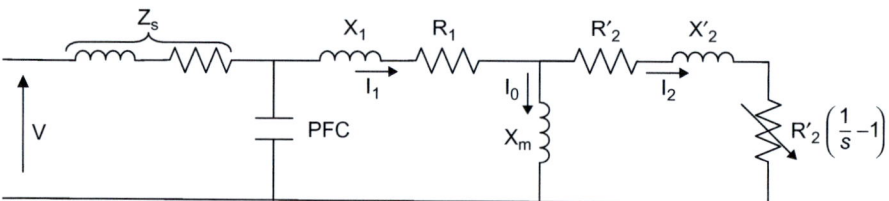

Figure 3.8   Steady-state equivalent circuit of induction machine connected through a source impedance (power factor correction (PFC) included)

As an example a group of ten of the 2 MW generators, as might be found in a wind farm, is considered. Each generator is compensated with 400 kVAr of power factor correction capacitors and connected through a 2 MVA generator transformer of 6% reactance to a busbar of short-circuit level of 200 MVA, which is represented by the source impedance connected to an infinite busbar. The group of ten generators is then considered as an equivalent single 20 MW generator (Figure 3.9). In the per unit system, this transformation is achieved conveniently by maintaining all the per-unit impedances of the generators, capacitors and transformers constant but

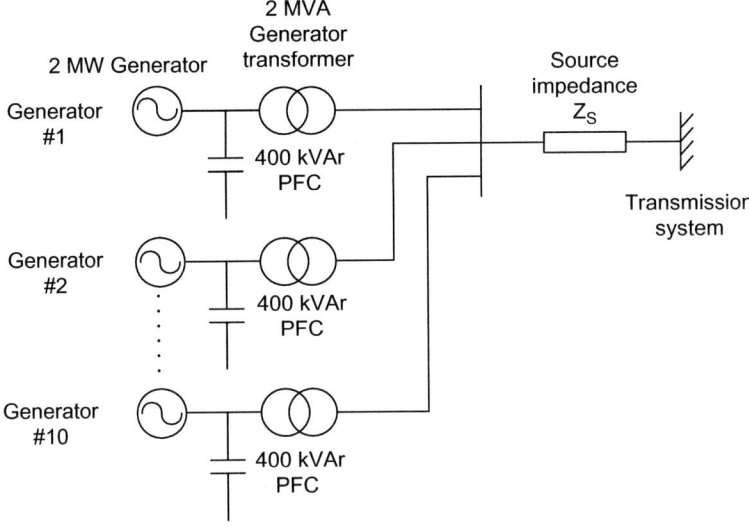

10 × 2 MW generators, equivalent circuit

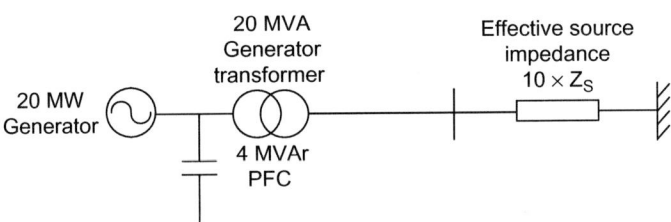

Figure 3.9 Representation of 10 × 2 MW coherent generators as a single 20 MW generator

merely changing the base MVA of the calculation. This has the effect of increasing the effective impedance of the connection to the infinite busbar by the number of generators (i.e. ten).

Figure 3.10 (left) shows the torque–slip curve of 10 and 30, 2 MW generators.[1] Increasing the number of turbines in the wind farm effectively increases the impact of the source impedance. The ten 2 MW turbines will operate satisfactorily on the 200 MVA fault-level connection but increasing the number of turbines to 30 is not possible.

With 30 turbines, the pull-out torque has dropped significantly to just below 1 pu (2 MW per turbine) due to the greater effect of the source impedance when the additional generators are connected. This would lead to instability with the generators no longer being able to transmit the torque applied by the prime mover.

---

[1] The 10 × 2 MW turbines are represented by a 20 MW coherent generator and results given on a 20 MVA base. Similarly the 30 × 2 MW turbines are represented by a 60 MW coherent generator with results given on a 60 MVA base.

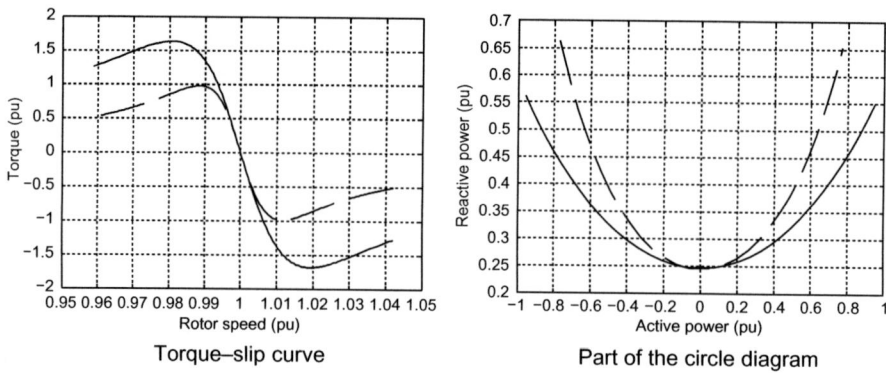

*Figure 3.10   Torque–slip curve and circle diagram of compensated 10 × 2 MW
            (solid) and 30 × 2 MW (dashed) coherent induction generators
            connected to a 200 MVA busbar*

The addition of the power factor correction capacitors translates the circle diagram
towards the origin, but the increase in number of connected machines demands more
reactive power at a given active power output, as shown in Figure 3.10 (right).

Figure 3.11 shows the variation of reactive power drawn with slip and it may be
seen that, if the 10 × 2 MW turbines all accelerated beyond their pull-out torque (at
around 2% slip) then some 40 MVAr (2 pu × 20 MVA base) would be demanded
from the network. This will clearly lead to voltage collapse in the network although,
in practice, the generators would have tripped on over-speed or under voltage.

This steady-state stability limit might be thought of as analogous to that of a
synchronous generator (II.6). If the steady-state stability limit of a synchronous

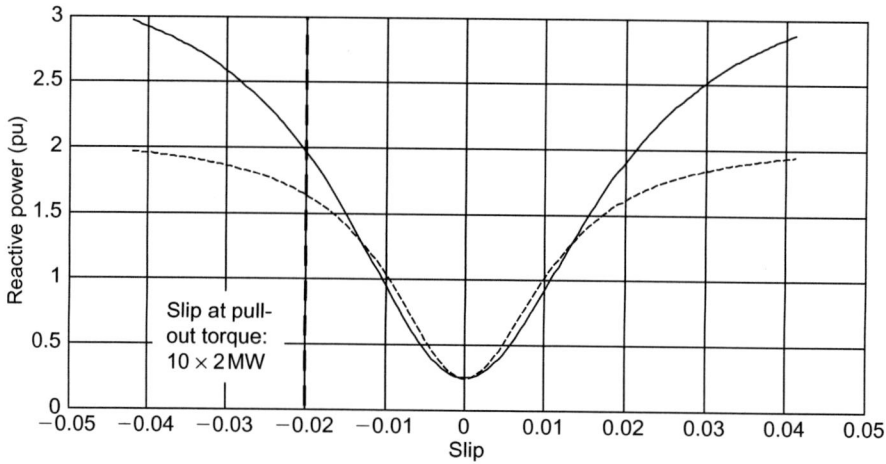

*Figure 3.11   Variation of reactive power drawn from network with slip (10 ×
            2 MW coherent induction generators – solid and 30 × 2 MW
            coherent induction generators – dotted)*

generator is exceeded, the rotor angle increases beyond 90° and pole slipping occurs. With an induction generator, when the pull-out torque is exceeded, excess reactive power is drawn to collapse the voltage and the generator accelerates until the prime mover is tripped.

For large wind farms on weak networks, including large offshore installations which will be connected to sparsely populated coasts, this form of instability may become critical. The voltage change in a radial circuit with an export of active power and import of reactive power is often estimated approximately by:

$$\Delta V = \frac{(PR - XQ)}{V} \qquad (3.1)$$

See Section 3.3.1 for the derivation of this equation[2].

It is frequently found that the circuit X/R ratio happens to be roughly equivalent to the P/Q ratio of an induction generator at near-full output. Hence, it may occur that the magnitude of the voltage at the generator terminals changes only slightly with load (although the relative angle and network losses will increase significantly) and so this potential instability may not be indicated by abnormal steady-state voltages.

A power-flow program with good induction machine steady-state models will indicate this potential voltage instability and fail to converge if the source impedance is too high for the generation proposed, although it is important to recognise that the governors on some distributed generator prime movers may not be precise and so operation at more than nominal output power must be investigated. The phenomena can be investigated more accurately using an electromagnetic transient or transient stability program with complete transient induction machine models or using one of the more recently developed continuation power-flow programs that can be used to find the actual point of voltage instability. A large induction generator, or collection of induction generators, which are close to their steady-state stability limit will, of course, be more susceptible to transient instability caused by network faults depressing the voltage.

Control over the power factor of the output of an induction generator is only possible by adding external equipment and the normal method is to add power factor correction capacitors at the terminals as shown in Figure 3.6. If sufficient capacitors are added then all the reactive power requirement of the generator can be supplied locally, and if connection to the network is lost, then the generator will continue to develop a voltage. In terms of distributed generation plant this is a most undesirable operating condition as, depending on the saturation characteristics of the induction generator, very large and distorted voltages can be developed as the generator accelerates. This phenomenon of 'self-excitation' has been reported as causing damage to load equipment connected to the isolated part of a network fed from induction generators with power factor correction.

---

[2] Note that as reactive power is imported by the induction generator, Q is negative with respect to the convention used in Section 3.3.1.

If the connection to the network is lost, the slip is small and as the stator leakage reactance and resistance are much less than the magnetising reactance, so the equivalent circuit of Figure 3.6 can be reduced to that of Figure 3.12 [2,8]. The magnetising reactance is shown as variable as its value changes with current due to magnetic saturation. Figure 3.12 represents a parallel resonant circuit with its operating points given by the intersection of the reactance characteristics of the capacitors (the straight lines) and the magnetising reactance that saturates at high currents. Thus, at frequency f_1 the circuit will operate at 'a' while as the frequency (rotational speed of the generator) increases to f_2 the voltage will rise to point 'b'. It may be seen that the voltage rise is limited only by the saturation characteristic of the magnetising reactance. Self-excitation may be avoided by restricting the size of the power factor correction capacitor bank to less than that required to make the circuit resonant at any credible generator speed (frequency) while its effect can be controlled by applying fast acting over-voltage protection on the induction generator circuit. Many presently available power system analysis programs do not include a representation of saturation in their induction machine models and so cannot be used to investigate this effect. In the detailed models found in electro-magnetic programs, saturation can be included if data is available. However, as self-excitation, and indeed any form of 'islanded' operation of induction generators, is a condition generally to be avoided, detailed investigation is not necessary for most distributed generation schemes.

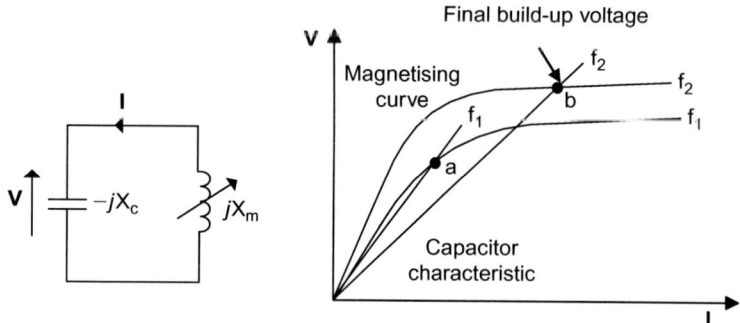

Figure 3.12   Representation of self-excitation of an induction generator

3.2.3   Doubly fed induction generator

A wound rotor induction machine can be operated as a variable-speed generator when a voltage is injected into the rotor circuit by an external means. A commonly used approach is the doubly fed induction generator system (see Figure 3.13), where the rotor converter controls the voltage of the wound rotor induction machine and hence its speed. The network side converter of the rotor circuit exchanges real power with the network and maintains the DC voltage across the capacitor. DFIG systems are now becoming widespread on large wind turbines.

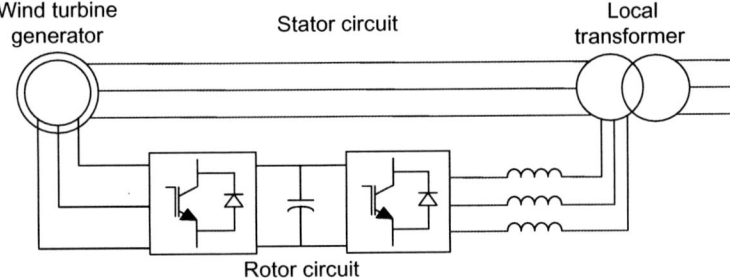

Figure 3.13 Doubly fed variable-speed power electronic converter (DFIG)

In wind turbine applications, the speed of the DFIG is controlled to extract maximum power from the wind. The power that can be extracted from the wind depends on the area swept by the turbine blades (A) and the wind speed (U) and is given by $P = (1/2)C_p\rho AU^3$, where ρ is the air density and C_p is the power coefficient [10]. The power coefficient depends on the tip speed ratio (λ) that is the ratio between the velocity of the rotor tip and wind speed. Therefore for a given wind speed, in order to extract maximum power the rotor speed of the generator should be varied. To control the generator speed, the controller shown in Figure 3.14 is used. The generator control is based on a dq coordinate system, where the q component of the stator voltage is selected as the real part of the busbar voltage and d component is the imaginary part [11,12]. This coordinate system allows the speed control action to be performed by manipulating the q component of the rotor-injected voltage, V_{qr}. The d component is controlled for the power factor and/or voltage control. More details of the controllers employed for DFIGs can be found in References 9 and 13.

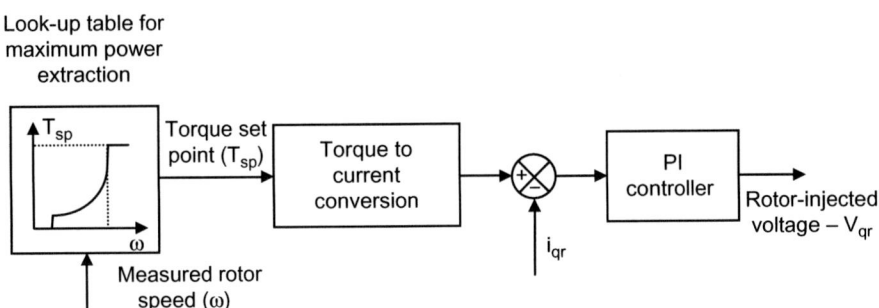

Figure 3.14 Torque controller for maximum power extraction

For a given injected voltage $\mathbf{V_r} = V_{dr} + jV_{qr}$, the performance of the DFIG wind turbine in steady state was obtained using the equivalent circuit shown in Figure 3.15.

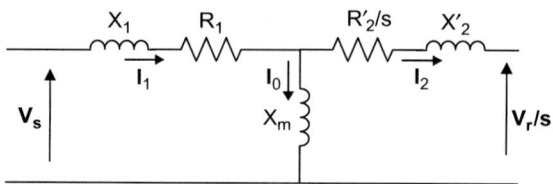

Figure 3.15   Steady-state equivalent circuit of the DFIG

Figure 3.16 shows the torque-slip curve of a 2 MW generator with three different rotor-injected voltages. The torque-slip characteristic for maximum power extraction is also shown in the figure. When the wind speed is high (hence the torque demanded by the speed control system shown in Figure 3.14 is high), the machine operates in super-synchronous speeds (point A of Figure 3.16). For lower wind speeds the machine operates in sub-synchronous speeds (point B of Figure 3.16).

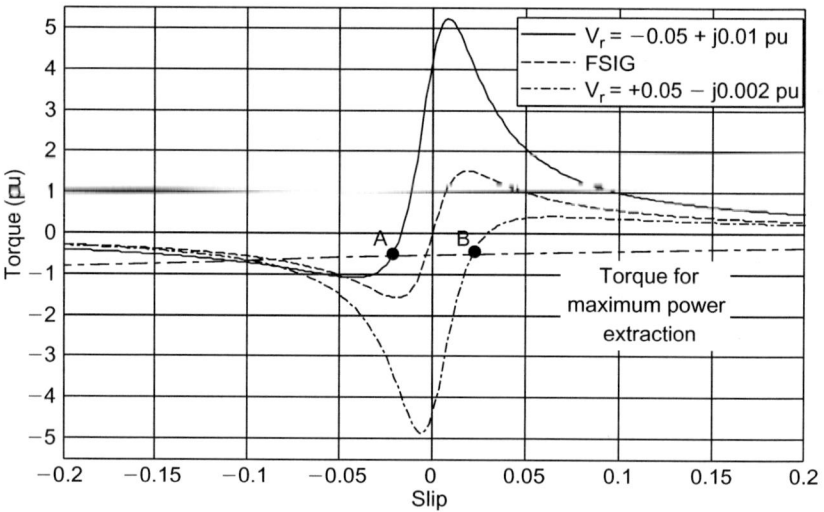

*Figure 3.16   Torque–slip curve for different rotor injections when a 2 MW DFIG
is connected to a 200 MVA busbar*

In a large wind farm, each DFIG wind turbine will be subjected to different wind speeds and the rotor-injected voltages that will be determined by the maximum power extraction controller are different. The torque–speed characteristic looks very different from machine to machine. Therefore, the analysis of a network with a DFIG wind farm may require detailed modelling of a number of generators. In contrast, for fixed speed induction generator based wind turbines the rotor-injected voltages are zero, as the rotors are all short circuited. Thus the torque–speed characteristics of all the machines are approximately the same. This allows the representation of a number of generators by a single coherent machine.

The DFIG wind turbine can be represented by a voltage behind a reactance in a similar manner to a synchronous generator. The phasor diagram of the DFIG (lagging operation, exporting VArs) is shown in Figure 3.17 [9]. In that diagram $\mathbf{E_g}$ is the voltage behind the reactance; $\mathbf{I_s}$, $\mathbf{V_s}$ and $\mathbf{V_r}$ are as defined in Figure 3.15 and X is given by: $X = X_1 + (X_m X'_2)/(X_m + X'_2)$.

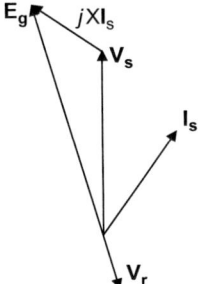

Figure 3.17   Phasor diagram of the DFIG

The phasor diagram for the stator is very similar to a synchronous generator, and the capability curve for the stator can be obtained in the same way as for a synchronous generator (Section II.2.4, Figure II.10). The main difference comes from the rotor injection. The total active power generation of a DFIG wind turbine is the addition of power through the stator (P_s) and power through the rotor (P_r). The following equations were derived using the generator torque, T and ignoring losses:

$$P_m = P_s + P_r$$
$$T\omega_r = T\omega_s + P_r \tag{3.2}$$
$$P_r = T(\omega_r - \omega_s) = -sT\omega_s = -sP_s$$

where ω_r is the rotor speed and ω_s is the synchronous speed.

From (3.2), it is clear that the rotor generates power during super-synchronous operation (where slip is negative) and it absorbs power during sub-synchronous operation. Therefore, the active power generation capability of a DFIG depends on the capability of power electronic converters and capability of stator conductors (heating limit).

The reactive power output of the DFIG is also controlled by the rotor-injected voltage. Figure 3.18 shows the reactive power as a function of active power for different rotor-injected voltages obtained using the DFIG equivalent circuit (Figure 3.15). The large number of possible operating points shown in the figure were taken by changing the rotor-injected voltage from $-0.05 + j0.05$ to $0.05 + j0.15$ and limiting the stator apparent power to 1.0 pu and the rotor apparent power to 0.3 pu.

The reactive power generation and absorption capability of the DFIG reduces with the active power. In wind generation applications this often requires the wind farm operator to connect a reactive power compensation device (e.g. a STATCOM) at the point of connection as otherwise it is not possible to fulfil the Grid Code requirements for reactive power.

EXAMPLE 3.1

For the wind farm shown in Figure E3.1, discuss how to achieve the Grid Code reactive power requirements, shown in Figure E3.2. Assume that each DFIG wind turbine has the capability curve shown in Figure 3.18.

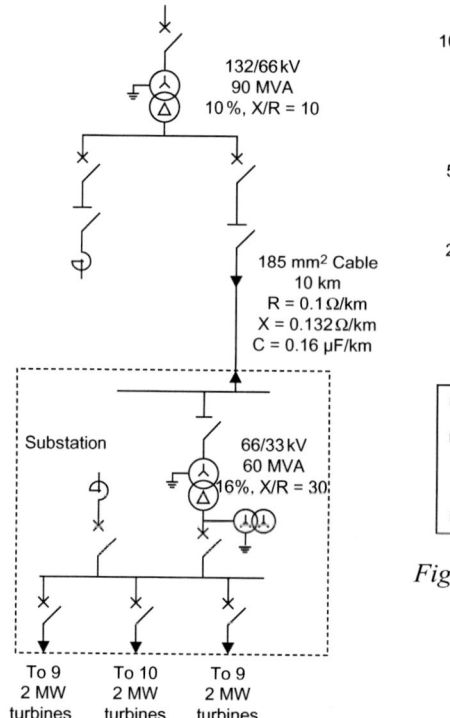

Point A is equivalent (in MVAr) to: 0.95 leading power factor at rated MW output

Point B is equivalent (in MVAr) to: 0.95 lagging power factor at rated MW output

Point C is equivalent (in MVAr) to: -5% of rated MW output

Point D is equivalent (in MVAr) to: +5% of rated MW output

Point E is equivalent (in MVAr) to: -12% of rated MW output

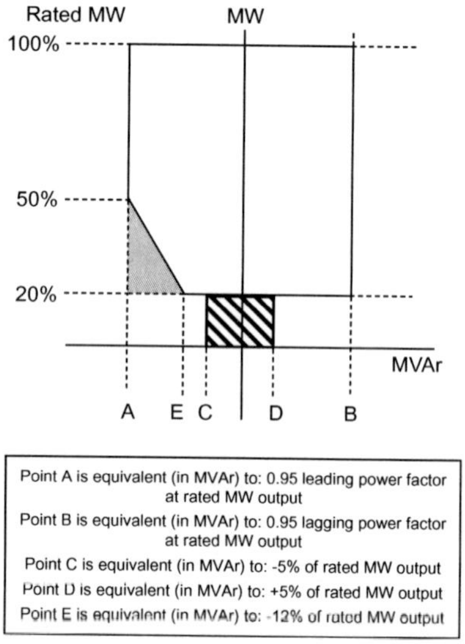

Figure E3.2   Great Britain reactive power requirement [14]

132/66 kV
90 MVA
10%, X/R = 10

185 mm² Cable
10 km
R = 0.1 Ω/km
X = 0.132 Ω/km
C = 0.16 μF/km

Substation

66/33 kV
60 MVA
16%, X/R = 30

To 9
2 MW
turbines

To 10
2 MW
turbines

To 9
2 MW
turbines

Distance between turbines 1km
185 mm² Cable
R = 0.13 Ω/km
X = 0.11 Ω/km

Figure E3.1   Example of wind farm network

The system shown in Figure E3.1 was simulated using the load flow of the IPSA computer package. Figure E3.3 shows the active and reactive power at the point of connection of the wind farm for different turbine operating points. The solid lines show the Grid Code reactive power requirements (Figure E3.2 translated for the wind farm). The two dotted curves are for 0.95 leading and lagging operation of all wind turbines. Points A and B are for 100% and 90% of power output respectively with the maximum reactive power capability of all wind turbines. The shaded area, bounded by curve AB and the Grid Code requirement, indicates an area where the wind farm cannot fulfil the Grid Code requirements.

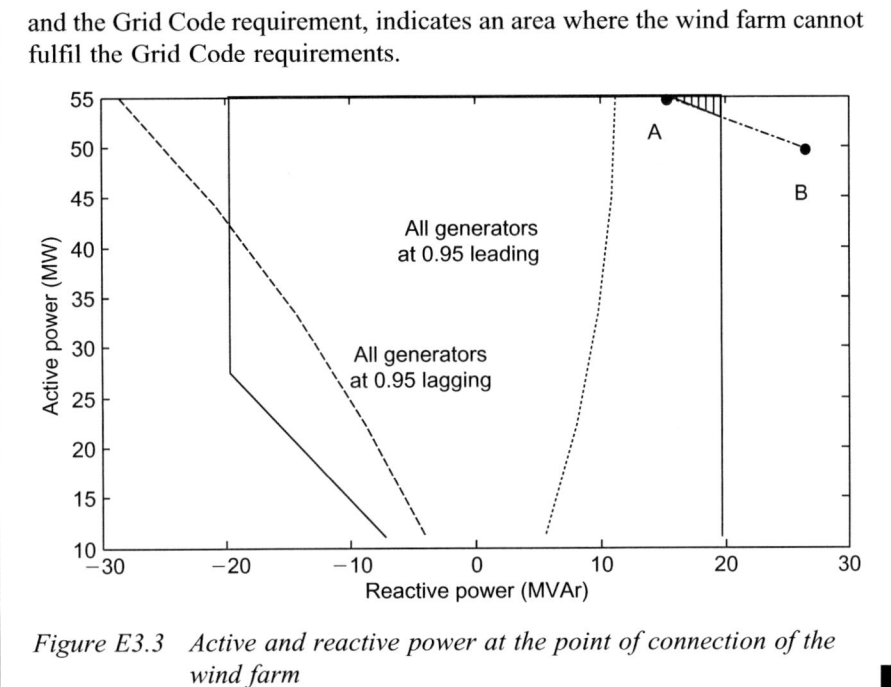

Figure E3.3   *Active and reactive power at the point of connection of the wind farm*

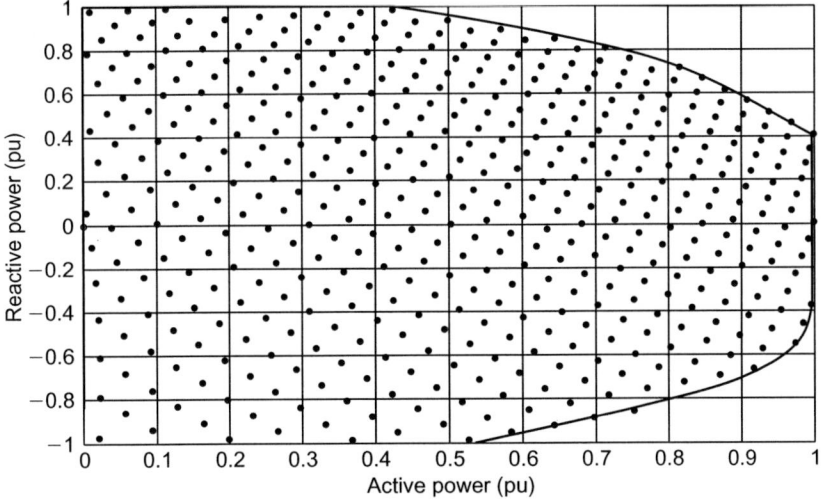

Figure 3.18   *Capability chart of the DFIG*

3.2.4   Full power converter (FPC) connected generators

Many renewable energy distributed generators use full power electronic converters to interface them to the network. The purpose of the power electronic interface depends on the application. For example in a PV system, the converter is used to invert DC generated by the PV modules to AC. In variable-speed wind turbines, back-to-back converters are used to extract maximum power from wind. Figure 3.19 shows a converter system typically used to control a large full power converter variable-speed wind turbine. The generator may be synchronous (wound rotor or permanent magnet) or an induction machine. Operation is possible over a wide speed range as all the power is rectified to DC and flows through the converters. Therefore, with typical losses of 2–3% in each converter it may be seen that, at full load, some 4–5% of the output power of the generator may be lost.

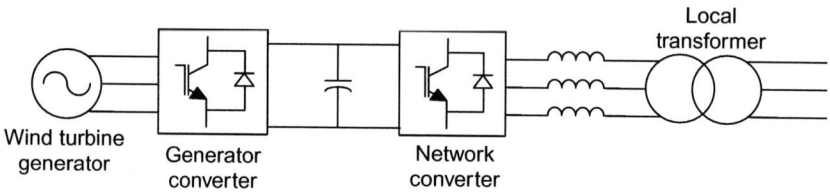

Figure 3.19   Full power converter (FPC) variable-speed generator

For large (>400 MW) offshore wind farms a long way offshore (>100 km), it may be cost effective to use HVDC (high voltage DC). HVDC can use one of the two technologies, current source converter (CSC) and voltage source converter (VSC), as shown in Figure 3.20. The CSC-based HVDC schemes are preferred for applications where power flow is very large (up to 3000 MW) and where there are synchronous generators to provide a commutating voltage at each end of the DC link. For distributed generation and for DC interfaces of moderate power wind farms, VSCs are becoming the preferred choice.

For assessing their impact on the power system, VSCs can be represented by a voltage behind a reactor as shown in Figure 3.21. This representation can be used for VSC HVDC as well as full power converter wind turbines and even some photo-voltaic systems.

Generally a phase locked loop (PLL) is employed to obtain the grid or gen-erator side voltage (busbar B) phase angle and frequency. A controller then turns ON and OFF the switches in the VSC to generate a voltage at busbar A with a phase angle relative to busbar B (depending on the control strategy). For system studies, neglecting higher-order harmonics generated by the VSC, this is represented by two voltage sources with reactive coupling between them.

With a VSC, sinusoidal current can only be injected if the IGBTs are switched rapidly. This leads to electrical losses, which may not be as commercially significant in a large motor drive as they will be in a distributed generation scheme. Thus, for large generators alternative arrangements may be considered including the use of multi-level inverters to combine a number of voltage sources or by combining mul-tiple inverters together through transformers of differing vector groups to form multi-phase inverters. The technique chosen will depend on an economic appraisal of the capital cost and the cost of losses, and as technology in this area is developing rapidly

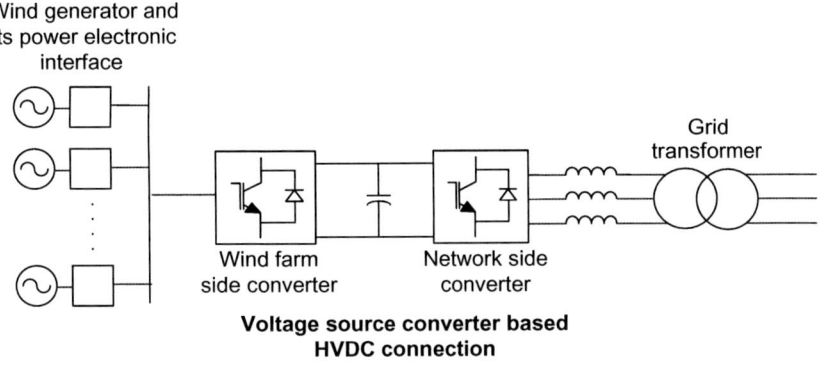

Voltage source converter based HVDC connection

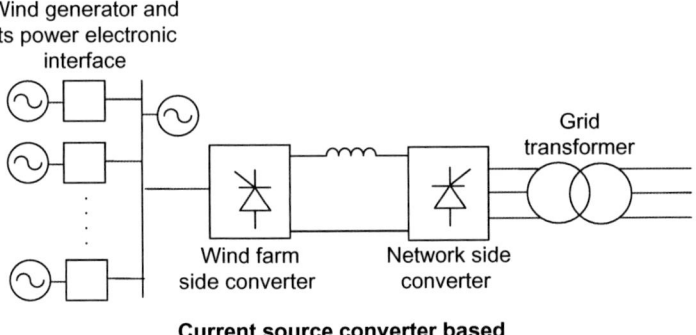

Current source converter based HVDC connection

Figure 3.20   Power electronic converter interfaces for large wind farms

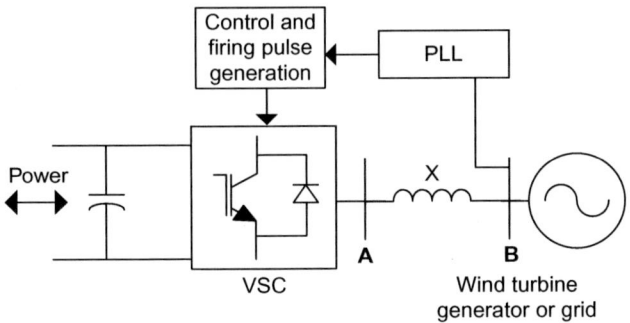

Figure 3.21   AC side connection of a VSC

so the most cost-effective technique at any rating will change over time. Future developments of power electronic converters for distributed generation plant may involve the use of soft-switching converters to reduce losses, resonant converters are already used in small photovoltaic generators, and include other converter topologies that eliminate the requirement for the explicit DC link.

3.3   System studies

3.3.1   *Load flow studies in a simple radial system*

Figure 3.22 shows the power flow of a two-busbar system.

From the expression for complex power, note that **S** is specified at the sending end busbar $\mathbf{S} = P + jQ = \mathbf{V_S}\mathbf{I}^*$:

$$\mathbf{I} = \frac{P - jQ}{\mathbf{V_S^*}} \tag{3.3}$$

The receiving end voltage is:

$$\mathbf{V_R} = \mathbf{V_S} - \mathbf{I}(R + jX) \tag{3.4}$$

Combining (3.3) and (3.4), and assuming the sending end voltage ($\mathbf{V_S}$) as the reference (i.e. $\mathbf{V_S} = V_S\angle 0°$):

$$\begin{aligned}\mathbf{V_R} &= \mathbf{V_S} - (R + jX)\left[\frac{P - jQ}{V_S}\right] \\ &= \mathbf{V_S} - \left[\frac{RP + XQ}{V_S}\right] - j\left[\frac{XP - RQ}{V_S}\right]\end{aligned} \tag{3.5}$$

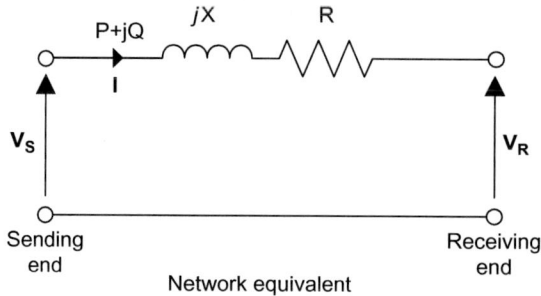

Network equivalent

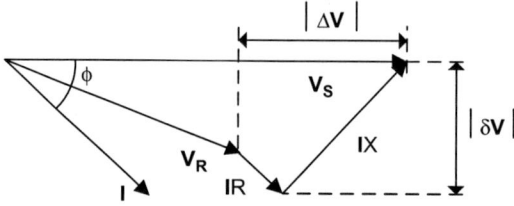

Phasor diagram

Figure 3.22   Two-busbar system

The right-hand side of (3.5) has sending end voltage, V_S, a component of voltage drop in phase with V_S and a component of voltage drop perpendicular to V_S. From the phasor diagram shown in Figure 3.22:

$$|\Delta V| = \left[\frac{RP + XQ}{V_S}\right] \tag{3.6}$$

$$|\delta V| = \left[\frac{XP - RQ}{V_S}\right] \tag{3.7}$$

For a distribution circuit $|\delta V|$ is normally neglected as the angle between V_R and V_S is very small. This approximation then allows a simple scalar approximate calculation of voltage drop (rise).

In a transmission circuit as $X \gg R$, the R terms in (3.6) and (3.7) are neglected.

These direct calculations are only possible if the sending end voltage and the sending end powers are known (so defining the current directly). Often the quantities known are the sending end voltage, V_S, and the receiving end active and reactive powers. This demands an iterative approach to find the receiving end voltage. If the receiving end active and reactive powers are P' and Q' respectively, then:

$$I = \frac{P' - jQ'}{V_R^*} \tag{3.8}$$

Now from (3.4):

$$V_R = V_S - (R + jX)\left[\frac{P' - jQ'}{V_R^*}\right] \tag{3.9}$$

The solution of (3.9) requires iteration.

Usually $V_R^{*(0)}$ is taken as 1 pu to start the iterative procedure.

Then $V_R^{(n)}$ is calculated and substituted into the right-hand side to determine $V_R^{(n+1)}$. This is continued until the values of V_R converge.

EXAMPLE 3.2

A 10 MW load of 0.9 power factor (pf) lagging (absorbing (VArs)) is connected to a 33 kV system through a cable of resistance 1.27 Ω and reactance 1.14 Ω. What is the receiving end voltage?

Answer:

Select: base MVA = 10 MVA and base voltage = 33 kV

Active power of the load: 10 MW = 1 pu

Since $\cos \phi = 0.9$, $\phi = 25.8°$

Reactive power of the load = $10 \times \tan(25.8°) = 4.8$ MVAr = 0.48 pu

Impedance base = $(33 \times 10^3)^2 / 10 \times 10^6 = 108.9\,\Omega$

Line resistance = 1.27/108.9 = 0.012 pu
Line reactance = 1.14/108.9 = 0.010 pu
From (3.9) with $V_R^{(0)} = 1$ pu

$$V_R^{(1)} = 1 - (0.012 + j0.01)\left[\frac{1 - j0.48}{1}\right]$$

$$= 0.983\angle 0.2°$$

Substituting $V_R^{(1)}$ again in (3.9):

$$V_R^{(2)} = 1 - (0.012 + j0.01)\left[\frac{1 - j0.48}{0.983\angle - 0.2°}\right]$$

$$= 0.982\angle - 0.23°$$

The above iterative procedure can be repeated until $|V_R^{(n+1)} - V_R^{(n)}| \leq \varepsilon$, where ε is a constant of convergence.

3.3.2   Load flow studies in meshed systems

Finding the voltage at a busbar using the two-bus method is not always possible as the network may be meshed. Consider the three bus system shown in Figure 3.23, where impedances, currents, load P and Q, and busbar voltages are marked. The known quantities are the generator terminal voltage, V_1 and the load powers $P_2 + jQ_2$ and $P_3 + jQ_3$. The two-bus method cannot be applied as it is not straightforward to calculate the current through the lines. Then algorithms based on iterative methods are required to solve for the unknown quantities. The programs that employ these algorithms are called load or power flows. There are many commercially available software packages used to perform load flow studies of large systems.

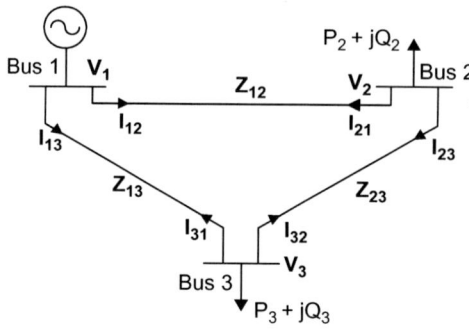

Figure 3.23   Three busbar power system

If the current from the generator is I_1, by applying Kirchhoff's current law to Bus 1:

$$I_1 = I_{12} + I_{13} = \frac{(V_1 - V_2)}{Z_{12}} + \frac{(V_1 - V_3)}{Z_{13}} \tag{3.10}$$

Defining the admittance as the reciprocal of the impedance (thus $Y_{12} = 1/Z_{12}$ and $Y_{13} = 1/Z_{13}$), (3.10) was rewritten as:

$$\begin{aligned} I_1 &= Y_{12}(V_1 - V_2) + Y_{13}(V_1 - V_3) \\ &= (Y_{12} + Y_{13})V_1 - Y_{12}V_2 - Y_{13}V_3 \end{aligned} \tag{3.11}$$

By applying Kirchhoff's current law to Bus 2 and using the corresponding admittances:

$$\begin{aligned} I_2 &= I_{21} + I_{23} \\ &= Y_{12}(V_2 - V_1) + Y_{23}(V_2 - V_3) \\ &= (Y_{12} + Y_{23})V_2 - Y_{12}V_1 - Y_{23}V_3 \end{aligned} \tag{3.12}$$

where I_2 is the negative value of the load current drawn by the load $P_2 + jQ_2$.
By applying Kirchhoff's current law to Bus 3:

$$\begin{aligned} I_3 &= I_{31} + I_{32} \\ &= Y_{13}(V_3 - V_1) + Y_{23}(V_3 - V_2) \\ &= (Y_{13} + Y_{23})V_3 - Y_{13}V_1 - Y_{23}V_2 \end{aligned} \tag{3.13}$$

where I_3 is the negative value of the load current drawn by the load $P_3 + jQ_3$.
Equations (3.11), (3.12) and (3.13) in matrix form:

$$\begin{bmatrix} I_1 \\ I_2 \\ I_3 \end{bmatrix} = \underbrace{\begin{bmatrix} (Y_{12} + Y_{13}) & -Y_{12} & -Y_{13} \\ -Y_{12} & (Y_{12} + Y_{23}) & -Y_{23} \\ -Y_{13} & -Y_{23} & (Y_{13} + Y_{23}) \end{bmatrix}}_{\text{Admittance matrix}=y} \begin{bmatrix} V_1 \\ V_2 \\ V_3 \end{bmatrix} \tag{3.14}$$

For a general case with N busbars:

$$I_k = \sum_{i=1}^{N} y_{ki}V_i = y_{kk}V_k + \sum_{\substack{i=1 \\ i \neq k}}^{N} y_{ki}V_i \tag{3.15}$$

where y_{ki} is the element in the kth row and ith column of the admittance matrix.

From (3.15):

$$\mathbf{V_k} = \frac{\mathbf{I_k}}{y_{kk}} - \frac{1}{y_{kk}} \sum_{\substack{i=1 \\ i \neq k}}^{N} y_{ki} \mathbf{V_i} \qquad (3.16)$$

For a load:

$$\mathbf{V_k} \mathbf{I_k^*} = -(P_k + jQ_k)$$

$$\mathbf{I_k} = \frac{-P_k + jQ_k}{\mathbf{V_k^*}} \qquad (3.17)$$

Combining (3.16) and (3.17):

$$\mathbf{V_k} = \frac{1}{y_{kk}} \left[\frac{-P_k + jQ_k}{\mathbf{V_k^*}} - \sum_{\substack{i=1 \\ i \neq k}}^{N} y_{ki} \mathbf{V_i} \right] \qquad (3.18)$$

For power-flow studies, one node is always specified as a voltage of constant magnitude and angle and is called the slack bus. For studies of distributed generation the slack bus is often a strong node of the main power system.

The voltage of a busbar to which a large synchronous generator is connected can be controlled by the generator excitation control. Therefore, such a busbar is specified as a generator bus, or a PV bus, where the magnitude of the voltage and active power are specified.

The third category of busbar is where a load is connected so that both active and reactive powers are specified (PQ bus).

An induction generator can be represented by a PQ bus with a negative active power and a positive reactive power as it generates P and absorbs Q. A distributed generator with a power electronic interface can be used for power factor and/or voltage control. If the voltage control is activated then it may be represented as a PV bus. If only power factor correction is activated then it can be represented by a PQ bus with a negative active power.

If there are N number of busbars in a power system, when one busbar is defined as the slack or reference bus there will be (N−1) simultaneous equations (3.18). The unknowns in these equations depend on the category of the busbar. For example for a generator bus, unknowns are the reactive power and voltage angle, whereas for a load bus the unknowns are the magnitude and angle of the voltage. Once the (N−1) equations are established, they can be solved using an iterative method. Two commonly used techniques are the: Gauss–Siedal method and Newton–Raphson method [4,5].

1.   Gauss–Siedal method

Choose an initial voltage for load busbars, say $\mathbf{V}_i^{(0)} = 1.0\angle 0°$ for $i = 1$ to N.

From (3.18):

$$\mathbf{V}_k^{(1)} = \frac{1}{y_{kk}} \left[\frac{-P_k + jQ_k}{(\mathbf{V}_k^{(0)})^*} - \sum_{\substack{i=1 \\ i \neq k}}^{N} y_{ki} \mathbf{V}_i^{(0)} \right]$$

Now solve for $\mathbf{V}_k^{(1)}$ for all the busbars. The iterative process can be accelerated if all the calculated values of $\mathbf{V}_k^{(1)}$ are used for subsequent calculations. For example when calculating $\mathbf{V}_3^{(1)}$, the calculated value of $\mathbf{V}_2^{(1)}$ is used.

The process is repeated until:

$$\left| \mathbf{V}_k^{(n+1)} - \mathbf{V}_k^{(n)} \right| \leq \varepsilon$$

where ε is a constant of convergence.

See Example 3.2 for details of the steps involved in this calculation method.

2. Newton–Raphson method

Equation (3.18) was applied to the three bus system shown in Figure 3.23. As busbar 1 is the slack bus (where the voltage is known), (3.18) was only applied to two PQ buses, i.e. busbars 2 and 3.

$$\mathbf{V}_2 = \frac{1}{y_{22}} \left[\frac{-P_2 + jQ_2}{\mathbf{V}_2^*} - \sum_{\substack{i=1 \\ i \neq 2}}^{3} y_{2i} \mathbf{V}_i \right] \tag{3.19}$$

$$\mathbf{V}_3 = \frac{1}{y_{33}} \left[\frac{-P_3 + jQ_3}{\mathbf{V}_3^*} - \sum_{\substack{i=1 \\ i \neq 3}}^{3} y_{3i} \mathbf{V}_i \right] \tag{3.20}$$

Equations (3.19) and (3.20) were rewritten as:

$$f_1(\mathbf{V}_2, \mathbf{V}_3) = C_1 \tag{3.21}$$
$$f_2(\mathbf{V}_2, \mathbf{V}_3) = C_2 \tag{3.22}$$

where $C_1 = y_{21} \mathbf{V}_1$ and $C_2 = y_{31} \mathbf{V}_1$ are constants.

$\mathbf{V}_2^{(0)}$ and $\mathbf{V}_3^{(0)}$ are the initial estimates of solutions to (3.21) and (3.22) and $\Delta \mathbf{V}_2^{(0)}$ and $\Delta \mathbf{V}_3^{(0)}$ are the values by which initial estimates differ from the correct solutions.

$$f_1(\mathbf{V}_2^{(0)} + \Delta \mathbf{V}_2^{(0)}, \mathbf{V}_3^{(0)} + \Delta \mathbf{V}_3^{(0)}) = C_1 \tag{3.23}$$
$$f_2(\mathbf{V}_2^{(0)} + \Delta \mathbf{V}_2^{(0)}, \mathbf{V}_3^{(0)} + \Delta \mathbf{V}_3^{(0)}) = C_2 \tag{3.24}$$

From Taylor's expansion neglecting higher-order derivates:

$$f_1(\mathbf{V}_2^{(0)}, \mathbf{V}_3^{(0)}) + \Delta\mathbf{V}_2^{(0)} \frac{\partial f_1}{\partial\mathbf{V}_2}\bigg|_{\mathbf{V}_2^{(0)}} + \Delta\mathbf{V}_3^{(0)} \frac{\partial f_1}{\partial\mathbf{V}_3}\bigg|_{\mathbf{V}_3^{(0)}} = C_1 \tag{3.25}$$

$$f_2(\mathbf{V}_2^{(0)}, \mathbf{V}_3^{(0)}) + \Delta\mathbf{V}_2^{(0)} \frac{\partial f_2}{\partial\mathbf{V}_2}\bigg|_{\mathbf{V}_2^{(0)}} + \Delta\mathbf{V}_3^{(0)} \frac{\partial f_2}{\partial\mathbf{V}_3}\bigg|_{\mathbf{V}_3^{(0)}} = C_2 \tag{3.26}$$

In matrix form:

$$\begin{bmatrix} C_1 - f_1(\mathbf{V}_2^{(0)}, \mathbf{V}_3^{(0)}) \\ C_2 - f_2(\mathbf{V}_2^{(0)}, \mathbf{V}_3^{(0)}) \end{bmatrix} = \underbrace{\begin{bmatrix} \frac{\partial f_1}{\partial\mathbf{V}_2} & \frac{\partial f_1}{\partial\mathbf{V}_3} \\ \frac{\partial f_2}{\partial\mathbf{V}_2} & \frac{\partial f_2}{\partial\mathbf{V}_3} \end{bmatrix}}_{\text{Jacobian matrix [4,5]}}_{\mathbf{V}_2^{(0)}, \mathbf{V}_3^{(0)}} \begin{bmatrix} \Delta\mathbf{V}_2^{(0)} \\ \Delta\mathbf{V}_3^{(0)} \end{bmatrix} \tag{3.27}$$

Equation (3.27) is solved to obtain $\Delta\mathbf{V}_2^{(0)}$ and $\Delta\mathbf{V}_3^{(0)}$. Then $\mathbf{V}_2$ and $\mathbf{V}_3$ were updated as:

$$\mathbf{V}_2^{(1)} = \mathbf{V}_2^{(0)} + \Delta\mathbf{V}_2^{(0)}$$
$$\mathbf{V}_3^{(1)} = \mathbf{V}_3^{(0)} + \Delta\mathbf{V}_3^{(0)}$$

The same procedure was repeated until $|\mathbf{V}_k^{(n+1)} - \mathbf{V}_k^{(n)}| \leq \varepsilon$, where ε is a constant of convergence.

For a large meshed network where there are PV buses and PQ buses, the formation of Jacobian matrix is more complex and more details can be found in References 4 and 5.

EXAMPLE 3.3

A large generator, G1,[3] is connected to Bus 1 of Figure E3.4 and maintains the voltage of that bus at $1.1\angle 0°$. Two loads connected to Bus 2 and 4 are $1 + j0.5$ and $0.5 + j0.25$ pu respectively. DG1 (a distributed generator) generates active power of 0.5 pu and absorbs reactive power of 0.2 pu. All the per unit quantities are on a 10 MVA base. Use the Gauss–Seidel method to determine the busbar voltages.

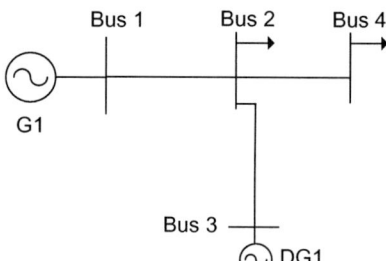

From bus	To bus	R (pu)	X (pu)
1	2	0.02	0.04
2	4	0.01	0.02
2	3	0.01	0.02

Figure E3.4

---

[3] In a study of Distributed Generation G1 represents the main power system.

Since $\mathbf{Z_{12}} = 0.02 + j0.04$ pu, $\mathbf{Y_{12}} = 1/(0.02 + j0.04) = 1/(0.0447\angle63.4°) = 22.36\angle-63.4° = 10.0 - j20.0$

Similarly: $\mathbf{Y_{23}} = 1/(0.01 + j0.02) = 20.0 - j40.0$ and $\mathbf{Y_{24}} = 1/(0.01 + j0.02) = 20.0 - j40.0$

Therefore the admittance matrix

$$= \begin{bmatrix} 10 - 20j & -10 + 20j & 0 & 0 \\ -10 + 20j & 50 - 100j & -20 + 40j & -20 + 40j \\ 0 & -20 + 40j & 20 - 40j & 0 \\ 0 & -20 + 40j & 0 & 20 - 40j \end{bmatrix}$$

Bus 1 is chosen as the slack bus. Therefore, (3.18) was applied to Buses 2, 3 and 4.

For Bus 2:

$$\mathbf{V_2} = \frac{1}{y_{22}}\left[\frac{-P_2 + jQ_2}{\mathbf{V_2^*}} - \sum_{\substack{i=1 \\ i\neq2}}^{4} y_{2i}\mathbf{V_i}\right]$$

$$= \frac{1}{50 - 100j}\left[\frac{-1 + j0.5}{\mathbf{V_2^*}} + (10 - 20j) \times 1.1\angle0° + (20 - 40j) \times \mathbf{V_3} + (20 - 40j) \times \mathbf{V_4}\right]$$

$$(3.28)$$

For an iterative solution, (3.28) was manipulated into:

$$\mathbf{V_2^{(1)}} = \frac{1}{(50 - 100j)}\left[11 - 22j + \frac{-1 + j0.5}{(\mathbf{V_2^{(0)}})^*} + (20 - 40j) \times \mathbf{V_3^{(0)}} + (20 - 40j) \times \mathbf{V_4^{(0)}}\right]$$

$$(3.29)$$

For Bus 3:[4]

$$\mathbf{V_3} = \frac{1}{y_{33}}\left[\frac{-P_3 + jQ_3}{\mathbf{V_3^*}} - \sum_{\substack{i=1 \\ i\neq3}}^{4} y_{3i}\mathbf{V_i}\right]$$

$$= \frac{1}{20 - 40j}\left[\frac{0.5 + j0.2}{\mathbf{V_3^*}} + (20 - 40j) \times \mathbf{V_2}\right]$$

---

[4] In (3.18), it was assumed that P and Q are flows from the busbar to the load. For the DG1, P flows towards the busbar and Q flows from the busbar.

Similarly to (3.29) but assuming that $V_2^{(1)}$ has already been calculated:

$$V_3^{(1)} = \frac{1}{(20 - 40j)} \left[\frac{0.5 + j0.2}{(V_3^{(0)})^*} + (20 - 40j) \times V_2^{(1)} \right] \quad (3.30)$$

For Bus 4:

$$V_4 = \frac{1}{y_{44}} \left[\frac{-P_4 + jQ_4}{V_4^*} - \sum_{\substack{i=1 \\ i \neq 4}}^{4} y_{4i} V_i \right]$$

$$= \frac{1}{20 - 40j} \left[\frac{-0.5 + j0.25}{V_4^*} + (20 - 40j) \times V_2 \right]$$

Similarly as (3.30):

$$V_4^{(1)} = \frac{1}{(20 - 40j)} \left[\frac{-0.5 + j0.25}{(V_4^{(0)})^*} + (20 - 40j) \times V_2^{(1)} \right] \quad (3.31)$$

To solve (3.29), (3.30) and (3.31) the Gauss-Siedal method is used. With $V_2^{(0)}$, $V_3^{(0)}$ and $V_4^{(0)}$ equal to $1.0 \angle 0°$, from (3.29):

$$V_2^{(1)} = \frac{1}{(50 - 100j)} \left[11 - 22j + \frac{-1 + j0.5}{1.0 \angle 0°} + (20 - 40j) \times 1.0 \angle 0° + (20 - 40j) \times 1.0 \angle 0° \right]$$

$$= 1.012 \angle -0.3°$$

With $V_2^{(1)} = 1.012 \angle -0.3°$ and $V_3^{(0)} = 1.0 \angle 0°$, from (3.30):

$$V_3^{(1)} = \frac{1}{(20 - 40j)} \left[\frac{0.5 + j0.2}{1.0 \angle 0°} + (20 - 40j) \times 1.012 \angle -0.3° \right]$$

$$= 1.013 \angle 0.3°$$

With $V_2^{(1)} = 1.012 \angle -0.3°$ and $V_4^{(0)} = 1.0 \angle 0°$, from (3.31):

$$V_4^{(1)} = \frac{1}{(20 - 40j)} \left[\frac{-0.5 + j0.25}{1.0 \angle 0°} + (20 - 40j) \times 1.012 \angle -0.3° \right]$$

$$= 1.002 \angle -0.8°$$

The following table shows the values of busbar 2, 3 and 4 voltages. The iterations were repeated until all voltages converged.

Iteration	V_2	V_3	V_4
1	$1.012\angle - 0.3°$	$1.013\angle 0.3°$	$1.002\angle - 0.8°$
2	$1.018\angle - 0.5°$	$1.019\angle 0.2°$	$1.008\angle - 0.9°$
3	$1.023\angle - 0.6°$	$1.024\angle 0°$	$1.013\angle - 1.0°$
4	$1.027\angle - 0.7°$	$1.028\angle - 0.1°$	$1.017\angle - 1.1°$
⋮	⋮	⋮	⋮
19	$1.043\angle - 1.0°$	$1.044\angle - 0.4°$	$1.033\angle - 1.4°$
20	$1.043\angle - 1.0°$	$1.044\angle - 0.4°$	$1.033\angle - 1.4°$

This example network was implemented in one of the commercially available load flow package (IPSA) and the voltages at busbars were obtained. The final voltages are $V_2 = 1.044\angle - 1.0°$, $V_3 = 1.045\angle - 0.4°$ and $V_4 = 1.034\angle - 1.4°$.

Example 3.3 illustrates that although the Gauss–Seidel method is simple in concept, it can be slow to converge for some networks and parameters. Thus all modern power-flow programs use the Newton–Raphson technique or a derivative of it.

3.3.3 Symmetrical fault studies

Fault calculations can be categorised by whether the currents and voltages are symmetrical or asymmetrical across the three phases. If a three-phase fault occurs then the network remains electrically balanced and the AC components of the resulting fault currents are symmetrical. A line-to-ground, line-to-line or line-to-line-to-ground fault results in asymmetrical fault currents.

Analysis of symmetrical faults is very similar to AC system analysis discussed in Tutorial Chapter IV [4,5,15]. Only one phase needs to be considered as the others are the same magnitude displaced by 120°.

EXAMPLE 3.4

In the circuit shown in Example IV.2 (in Tutorial Chapter IV), calculate the fault current resulting from a fault at the end of the line.

Answer:

The per unit equivalent circuit is given by:

Neglecting resistance,[5] the fault current $= 1/(0.15 + 0.1 + 4.59)$ $= 0.207$ pu.

With S_b 100 MVA and V_L 33 kV, the current base is:

$$I_b = \frac{S_b}{\sqrt{3}V_L} = \frac{100 \times 10^6}{\sqrt{3} \times 33 \times 10^3} = 1749.5 \text{ A}$$

Therefore, the fault current $= 1749.5 \times 0.207 = 361.5$ A.

EXAMPLE 3.5

The drawing shows the synchronous generator of a solar thermal power plant connected to a radial system.

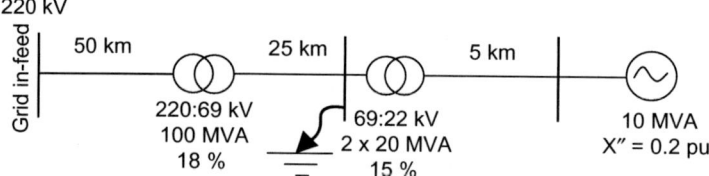

The parameters of the overhead lines are:

Nominal voltage of circuit	Impedance of line (Ω/km)
220 kV	j3.0
69 kV	0.5 + j1.0
22 kV	0.5 + j0.25

Using a 100 MVA base, change all the parameters into per unit. A three-phase short circuit fault occurs at the 69 kV busbar of the 69/22 kV transformer substation. Calculate the fault current in per unit and amperes that flows from the infinite busbar and the generator.

Answer:

X'' is the reactance of the generator at the moment a fault occurs. On 100 MVA base $X'' = 0.2 \times 100/10 = j2$ pu.

The following table gives the base impedance (Z_B) and the per unit reactance of each line, neglecting resistance.

---

[5] If all network resistances are neglected, the resulting fault currents will be slightly high and so the calculation is generally conservative. The j operator may be ignored.

Nominal voltage of circuit (kV)	Z_B (Ω)	Reactance of line (Ω)	Reactance of line (pu)
220	$\dfrac{(220 \times 10^3)^2}{100 \times 10^6} = 484$	$j150$	$j0.31$
69	$\dfrac{(69 \times 10^3)^2}{100 \times 10^6} = 47.6$	$j25$	$j0.525$
22	$\dfrac{(22 \times 10^3)^2}{100 \times 10^6} = 4.84$	$j1.25$	$j0.258$

220 kV:69 kV transformer reactance on 100 MVA base = $j0.18$ pu.

69 kV:22 kV transformer reactance on 100 MVA base = $j0.15 \times 100/40 = j0.375$.

Therefore, the per unit equivalent of the system may be drawn as (note that the circuit is rearranged to make the subsequent calculations clearer):

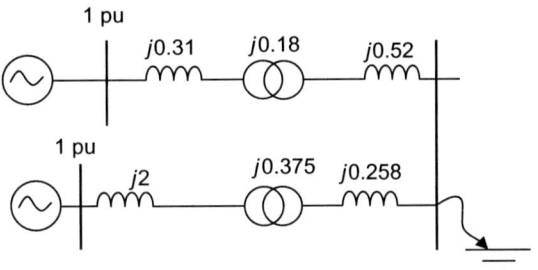

Fault current flowing from the infinite busbar = $1/(0.31 + 0.18 + 0.525)$ = 0.985 pu.

Fault current flowing from the generator = $1/(2 + 0.375 + 0.258) = 0.38$ pu.

The base current on 220 kV network is $100 \times 10^6 / \sqrt{3} \times 220 \times 10^3 = 262.4$ A.

Therefore, the fault current flowing from the infinite busbar = $0.985 \times 262.4 = 258.5$ A.

The base current on the 22 kV network is $100 \times 10^6 / \sqrt{3} \times 22 \times 10^3 = 2624.3$ A.

Therefore, the fault current flowing from the infinite busbar = $0.38 \times 2624.3 = 997.2$ A.

3.3.4 Unbalanced (asymmetrical) fault studies

The currents in each phase resulting from an unbalanced fault (line-ground, line-line or line-line-ground) are not equal and so are calculated using the method of symmetrical components, discussed in Section IV.4 [4,5,15]. When using symmetrical components, the power system is represented as three sequence networks: positive, negative and zero. The relationship between the phase voltages and sequence network voltages is given by (IV.33) and repeated here:[6]

$$\begin{bmatrix} \mathbf{V_{A0}} \\ \mathbf{V_{A1}} \\ \mathbf{V_{A2}} \end{bmatrix} = \frac{1}{3} \begin{bmatrix} 1 & 1 & 1 \\ 1 & \lambda & \lambda^2 \\ 1 & \lambda^2 & \lambda \end{bmatrix} \begin{bmatrix} \mathbf{V_A} \\ \mathbf{V_B} \\ \mathbf{V_C} \end{bmatrix} \tag{3.32}$$

To illustrate the formation of sequence networks, the line section shown in Figure 3.24 is used.

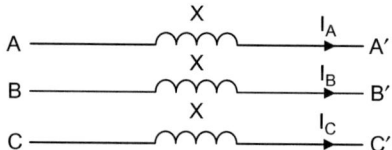

Figure 3.24   Line section with pure reactances

The voltage and currents for the line section are:

$$\begin{bmatrix} \mathbf{V_{AA'}} \\ \mathbf{V_{BB'}} \\ \mathbf{V_{CC'}} \end{bmatrix} = X \begin{bmatrix} \mathbf{I_A} \\ \mathbf{I_B} \\ \mathbf{I_C} \end{bmatrix} \tag{3.33}$$

Three-phase voltage drops given by (3.33) were converted to three sequence components by using (3.32):

$$\begin{bmatrix} \mathbf{V_{AA'0}} \\ \mathbf{V_{AA'1}} \\ \mathbf{V_{AA'2}} \end{bmatrix} = \frac{1}{3} \begin{bmatrix} 1 & 1 & 1 \\ 1 & \lambda & \lambda^2 \\ 1 & \lambda^2 & \lambda \end{bmatrix} \begin{bmatrix} \mathbf{V_{AA'}} \\ \mathbf{V_{BB'}} \\ \mathbf{V_{CC'}} \end{bmatrix} = \frac{1}{3} X \begin{bmatrix} 1 & 1 & 1 \\ 1 & \lambda & \lambda^2 \\ 1 & \lambda^2 & \lambda \end{bmatrix} \begin{bmatrix} \mathbf{I_A} \\ \mathbf{I_B} \\ \mathbf{I_C} \end{bmatrix} \tag{3.34}$$

As (3.32) is also true for currents:

$$\begin{bmatrix} \mathbf{I_{A0}} \\ \mathbf{I_{A1}} \\ \mathbf{I_{A2}} \end{bmatrix} = \frac{1}{3} \begin{bmatrix} 1 & 1 & 1 \\ 1 & \lambda & \lambda^2 \\ 1 & \lambda^2 & \lambda \end{bmatrix} \begin{bmatrix} \mathbf{I_A} \\ \mathbf{I_B} \\ \mathbf{I_C} \end{bmatrix} \tag{3.35}$$

From (3.34) and (3.35):

$$\begin{bmatrix} \mathbf{V_{AA'0}} \\ \mathbf{V_{AA'1}} \\ \mathbf{V_{AA'2}} \end{bmatrix} = X \begin{bmatrix} \mathbf{I_{A0}} \\ \mathbf{I_{A1}} \\ \mathbf{I_{A2}} \end{bmatrix} \tag{3.36}$$

---

[6] $\lambda = e^{j2\pi/3}$ or $120°$ phase shift.

From (3.36), it may be seen that each component voltage drop only depends on that component current and the reactance.

Figure 3.25 shows the cross section of a cable. As the three conductors are equally spaced, the cable may be considered a symmetrical element with similar distribution of magnetic flux around each conductor. For such a symmetrical non-rotating element, the positive and negative sequence inductances are equal.

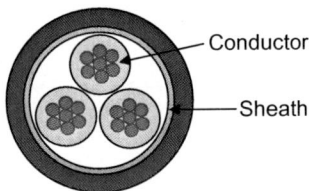

Figure 3.25   Cross section of a three-phase cable

The flux that links with the sheath (which is earthed) determines the zero sequence inductance, with a different value to the other two sequence components.

A section of an overhead line is shown in Figure 3.26. In this case flux linkages between different phases are not equal and therefore the positive and negative sequence inductances are different [5].

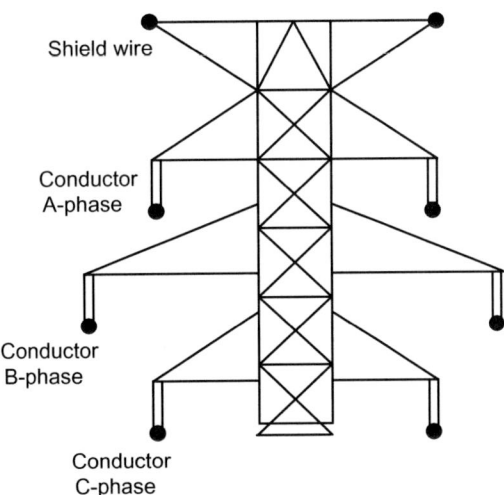

Figure 3.26   A section of an overhead line

The internal impedance of a generator, expressed in sequence values, depends on its construction. The magnetic field produced by the positive sequence currents rotates in the same direction as the rotor; thus with respect to the rotor it is stationary. The field produced by the negative sequence currents rotates in the opposite direction to the rotor. Therefore with respect to the rotor that magnetic field rotates at twice the rotor speed. These differences lead to different positive and negative sequence impedances.

Consider an ideal, balanced three-phase source shown in Figure 3.27 where neutral point, N, is earthed through an impedance Z_n.

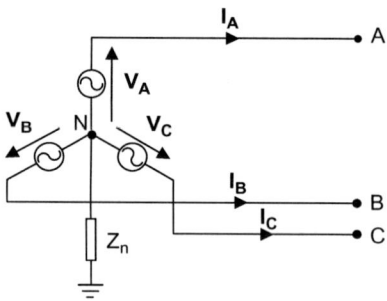

Figure 3.27   Three-phase source

The internal generated voltages of each phase are given by the following three equations where E is the rms value of the internal voltage and:

$$\left.\begin{array}{l} \mathbf{E_A} - E\angle0° - E \\ \mathbf{E_B} = E\angle-120° = E\angle240° = \lambda^2 E \\ \mathbf{E_C} = E\angle-240° = E\angle120° = \lambda E \end{array}\right\} \tag{3.37}$$

From (3.32) and (3.37) (note that from Figure IV.8 of the Tutorial Chapter IV $(1 + \lambda + \lambda^2) = 0$, $(1 + 2\lambda^3) = 3$ and $(1 + \lambda^2 + \lambda^4) = 0$):

$$\begin{bmatrix} \mathbf{V_{A0}} \\ \mathbf{V_{A1}} \\ \mathbf{V_{A2}} \end{bmatrix} = \frac{1}{3}\begin{bmatrix} 1 & 1 & 1 \\ 1 & \lambda & \lambda^2 \\ 1 & \lambda^2 & \lambda \end{bmatrix}\begin{bmatrix} E \\ \lambda^2 E \\ \lambda E \end{bmatrix} = \begin{bmatrix} 0 \\ E \\ 0 \end{bmatrix} \tag{3.38}$$

From (3.38), it is clear that the ideal three-phase source does not produce any negative or zero sequence voltages. Therefore, a three-phase source can be represented by three sequence networks where the positive sequence network has a voltage equal to $\mathbf{E_A} = E$ and the negative and zero sequence networks have zero voltages.

Three sequence voltages are represented behind the internal impedance of the source. With positive sequence impedance of Z_1, the positive sequence network with respect to the neutral was obtained as:

$$\mathbf{V_{A1}} = \mathbf{E_A} - \mathbf{I_{A1}}Z_1 \tag{3.39}$$

With negative sequence impedance of Z_2 the negative sequence network with respect to the neutral was obtained as:

$$V_{A2} = -I_{A2}Z_2 \tag{3.40}$$

Finally with zero sequence impedance of Z_0, the zero sequence network with respect to earth was obtained as:

$$V_{A0} = V_N - I_{A0}Z_0 \tag{3.41}$$

Since the impedance between the neutral and earth is Z_n, then:

$$V_N = -[I_A+I_B+I_C]Z_n \tag{3.42}$$

From (3.35), $[I_A+I_B+I_C] = 3I_{A0}$ and therefore from (3.42):

$$V_N = -3I_{A0}Z_n \tag{3.43}$$

By substituting for V_N from (3.43) into (3.41):

$$V_{A0} = -I_{A0}(Z_0 + 3Z_n) \tag{3.44}$$

Three sequence networks for a three-phase source are shown in Figure 3.28. If the source is delta connected or star connected without a neutral then $Z_0 \rightarrow \infty$.

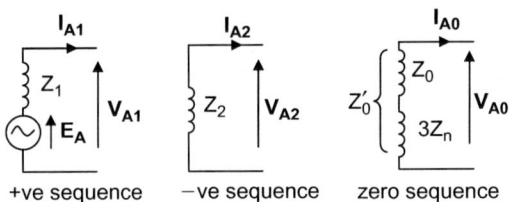

Figure 3.28   Sequence networks for a three-phase source

For a transformer, the positive and negative sequence impedances are equal. Depending on the connection used for the primary and secondary of the transformer, the zero sequence component current is transferred from the primary to secondary or is blocked (Table 3.1).

Once the sequence component networks of the power system are known, then the asymmetrical currents are obtained by connecting the networks in different arrangements. The connection of three sequence networks depends on the type of the fault:

A line-to-earth fault through a fault impedance, Z_f, is shown in Figure 3.29.

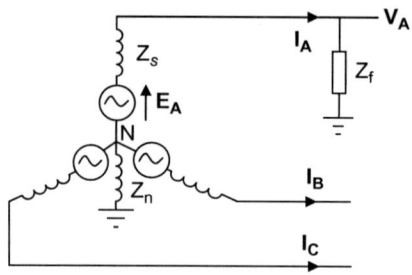

Figure 3.29   Line-to-earth fault

Table 3.1   Zero sequence component and transformer connection

Transformer connection		Sequence networks (see Figure IV.6)	
Primary	Secondary	Positive and negative	Zero
			Zero sequence current flows from the primary side to the secondary side.
			Zero sequence current enters the transformer from the primary side but will not flow to the secondary side as the unearthed star blocks the zero sequence current.
			Zero sequence current enters the transformer from the primary side but will not flow to the secondary side as the currents circulate in the delta loop.

In this case, the fault current in the B and C phases are zero and fault current only flows in A phase. From symmetrical components, when $\mathbf{I_B} = \mathbf{I_C} = 0$:

$$\begin{bmatrix} \mathbf{I_{A0}} \\ \mathbf{I_{A1}} \\ \mathbf{I_{A2}} \end{bmatrix} = \frac{1}{3} \begin{bmatrix} 1 & 1 & 1 \\ 1 & \lambda & \lambda^2 \\ 1 & \lambda^2 & \lambda \end{bmatrix} \begin{bmatrix} \mathbf{I_A} \\ 0 \\ 0 \end{bmatrix} \tag{3.45}$$

$$\therefore \mathbf{I_{A0}} = \mathbf{I_{A1}} = \mathbf{I_{A2}} = \frac{1}{3}\mathbf{I_A}$$

From Figure 3.29:

$$\mathbf{V_A} = \mathbf{I_A}Z_f \tag{3.46}$$

Now substituting for $\mathbf{V_A}$ and $\mathbf{I_A}$ with their symmetrical components (see (IV.26)):

$$\mathbf{V_{A1}} + \mathbf{V_{A2}} + \mathbf{V_{A0}} = [\mathbf{I_{A1}} + \mathbf{I_{A2}} + \mathbf{I_{A0}}]Z_f \tag{3.47}$$

Substituting from (3.45) into (3.47), the following equation is obtained:

$$\mathbf{V_{A1}} + \mathbf{V_{A2}} + \mathbf{V_{A0}} = 3\mathbf{I_{A0}}Z_f \tag{3.48}$$

It is clear that if the sequence networks are connected in series, as shown in Figure 3.30, then both (3.45) and (3.48) are satisfied.

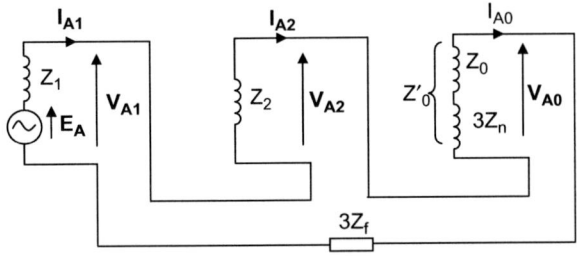

Figure 3.30   Connection of sequence networks for a line-to-earth fault

The formation of the sequence networks for other types of faults are given in Appendix (A3.1).

EXAMPLE 3.6

For the circuit shown, on 50 MVA base the sequence impedances of the components are:

	Positive sequence	Negative sequence	Zero sequence
Circuit	j0.6	j0.6	j1.5
Transformer	j0.1	j0.1	j0.1

A single phase to earth fault occurred at the mid point of the line. Find the fault current.

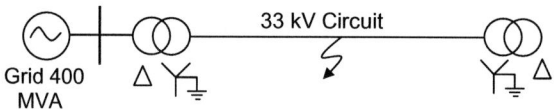

Grid 400 MVA   33 kV Circuit

Answer:

Positive sequence network:

  The fault level at the grid in-feed = 400/50 pu = 8 pu.

  Source impedance – 1/8 pu – 0.125 pu.

  Transformer impedance = 0.1 pu.

  Line impedance = 0.6/2 = 0.3 pu.

Therefore, the total positive sequence impedance to the fault:

$$Z_1 = j(0.125 + 0.1 + 0.3) = j0.525 \, \text{pu}$$

The positive sequence network is given by:

j0.525 pu

1 pu   Fault point

Negative sequence network:

  Transformer impedance = 0.1 pu

  Line impedance = 0.6/2 = 0.3 pu

The total negative sequence impedance to the fault:

$$Z_2 = j(0.1 + 0.3) = j0.4 \, \text{pu}$$

The negative sequence network is given by:

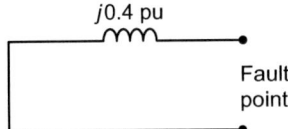

Zero sequence network:

For zero sequences, it is considered that the line is connected to the earth from both ends (due to star earth on the transformers). Therefore, the zero sequence network is given by:

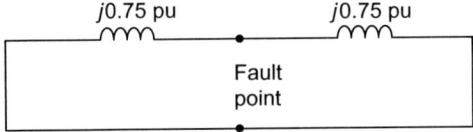

Therefore, the total negative sequence impedance to the fault:

$$Z_0 = j0.75/2 = j0.375 \, \text{pu}$$

For a single phase to earth fault, three sequence networks are connected in series. Assuming that the fault impedance $Z_f = 0$, the equivalent circuit for a line-to-earth fault is given by:

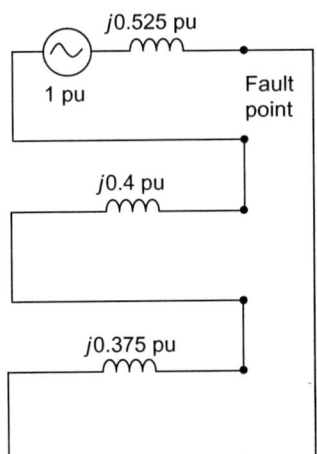

From the above figure and from (3.45):

$$\mathbf{I_{A0}} = \mathbf{I_{A1}} = \mathbf{I_{A2}} = \frac{1}{3}\mathbf{I_A} = \frac{1}{0.525 + 0.4 + 0.375} = 0.77 \, \text{pu}$$

Therefore, the fault current = 3 × 0.77 = 2.31 pu.
On 50 MVA, 33 kV (line) base, the current base is given by:

$$I_b = \frac{S_b}{3V_b} = \frac{50 \times 10^6}{3 \times (33 \times 10^3/\sqrt{3})} = 874.78 \text{ A}$$

Therefore, the fault current = 874.78 × 2.31 = 2020.7 A.

3.4   Case studies

3.4.1   Steady-state voltages under peak and minimum loading

A section of a weak 33 kV network is shown in Figure 3.31. The generator connected to Bus 1 represents the grid in-feed, which supplies power from the 132 kV
transmission system to the 33 kV distribution network. The fault level at the grid in-
feed is 1850 MVA. The generator, transformer, line and load data are given in
Tables 3.2–3.5. The 132/33 kV transformer has an on-load tap changer which
maintains Bus 2 at a voltage at 1.005 pu. The tap changer at the distributed generator (DG) is off-load and set to 0%.

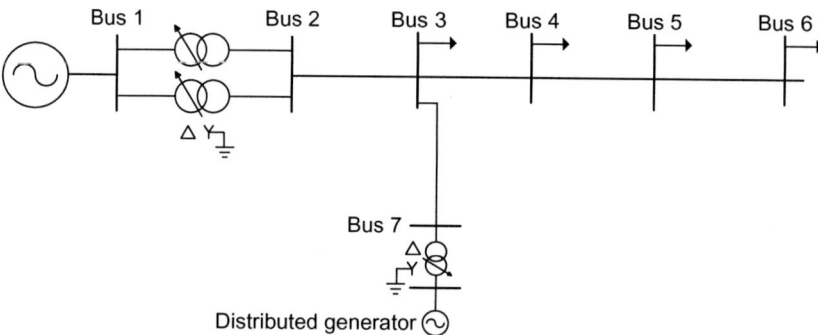

Figure 3.31   33 kV distribution system with a distributed generator

Table 3.2   Distributed generator data

Name	MVA rating	X_d (pu)	X_d' (pu)	X_q (pu)	T_d (s)
DG	8	0.29	0.16	0.29	4.5

Table 3.3   Transformer data

From bus	To bus	MVA rating	Winding	Reactance (%)
Bus 1	Bus 2	31.50	Yd1	10.5
Bus 7	DG	11.00	Yd1	8.8

Table 3.4   Circuit data

From bus	To bus	Length (km)	R (pu/km)	X (pu/km)
Bus 2	Bus 3	17.2		
Bus 3	Bus 4	12.0		
Bus 3	Bus 7	0.5	0.016250	0.034435
Bus 4	Bus 5	0.8		
Bus 5	Bus 6	1.4		

Table 3.5   Load data (all loads have a lagging power factor)

Busbar	Peak		Minimum	
	P (MW)	Q (MVAr)	P (MW)	Q (MVAr)
3	2.0	0.48	0.25	0.06
4	2.0	0.48	0.25	0.06
5	2.4	0.56	0.30	0.07
6	3.2	0.80	0.40	0.10

1.   Steady-state voltages, power flows and losses

A load flow study was carried out using the IPSA package and voltages; power flows and losses were obtained for peak and minimum loading conditions. Figure 3.32 shows the load flow under peak and minimum loading conditions when DG is importing VArs (0.95 leading power factor). It is clear from the diagram that during the peak loading condition power flows from upstream to downstream via the line between Buses 2 and 3. However, during the minimum loading condition excess power in the distribution network is fed back to the grid.

Table 3.6 shows the voltage on busbars 3, 4, 5 and 6 and losses when a distributed generator is not connected, when it is operating at unity, 0.95 lagging (exporting VArs[7]) and 0.95 leading (importing VArs) power factors. As the network is weak (with a high impedance), under peak loading condition the voltage at some busbars are below the limit (±6%) when the distributed generator is not connected. The distributed generator reduced the power flow in the line between Bus 2 and 3, thus reducing the losses. When the distributed generator operates at unity power

[7] Note that for a load, lagging power factor means it absorbs reactive power.

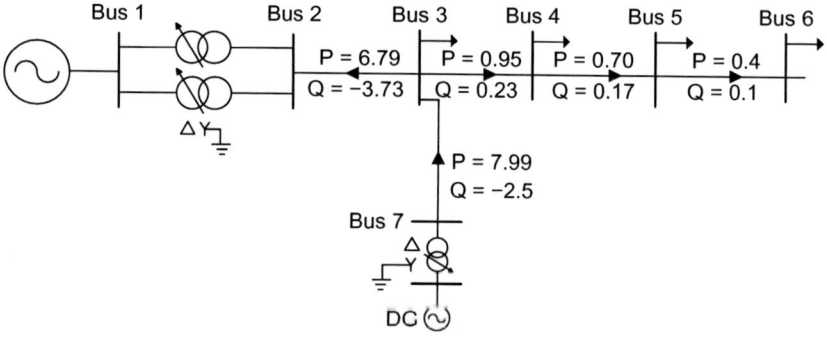

Figure 3.32   *Power flows in peak and minimum loading conditions (P in MW and Q in MVAr)*

Table 3.6   *8 MVA distributed generator is connected to Bus 3*

	Voltage (pu)				Losses	
	Bus 3	Bus 4	Bus 5	Bus 6	MW	MVAr*
Peak loading						
No distributed generator	0.959	0.934	0.933	0.932	0.46	1.15
Distributed generator at unity pf	0.981	0.957	0.956	0.955	0.18	0.94
Distributed generator at 0.95 pf leading	0.964	0.940	0.939	0.938	0.26	1.25
Distributed generator at 0.95 pf lagging	0.996	0.973	0.972	0.970	0.15	0.88
Minimum loading						
No distributed generator	1.000	0.997	0.997	0.997	0.01	0.02
Distributed generator at unity pf	1.018	1.015	1.015	1.015	0.13	0.85
Distributed generator at 0.95 pf leading	1.003	1.000	1.000	1.000	0.17	1.04
Distributed generator at 0.95 pf lagging	1.032	1.030	1.030	1.030	0.14	0.87

*Reactive power losses are the reactive power absorbed by the circuits.

factor, the voltage profile is improved. The voltage profile can be improved further by operating the distributed generator at 0.95 lagging pf (exporting VArs). The converse is true when the distributed generator operates at a leading power factor (importing VArs). Under minimum loading conditions the case without the distributed generator shows minimum losses and a good voltage profile. The connection of the distributed generator increases the voltages at the busbars and excess power is now transmitted back to the grid thus increasing the losses.

2.   Effect on load compensation

As shown in Table 3.6, the voltages at remote busbars when the distributed generator is not in operation are below the limit. It is customary to use line drop compensation (LDC) on the automatic voltage controller (AVC) of the on-load tap changing (OLTC) transformer to address this [16]. The LDC controls the voltage at a remote point downstream of the OLTC transformer. One concern utilities have over distributed generator connection is possible interaction with LDC when the distributed generator feeds power back to the grid. This is a reversal of the conventional power flow to which the AVC with LDC is designed to respond.

   This case was simulated by designing the AVC with LDC to maintain the voltage at Bus 6 at 0.95 pu. Figure 3.33 shows the voltages at each busbar for peak loading condition without the distributed generator and minimum loading condition with the DG (operating at 0.95 lagging pf). As can be seen from the top diagram of Figure 3.33, the voltage at Bus 6 has increased from 0.932 to 0.953, thus showing the expected LDC operation. However, when the DG is connected the voltage at

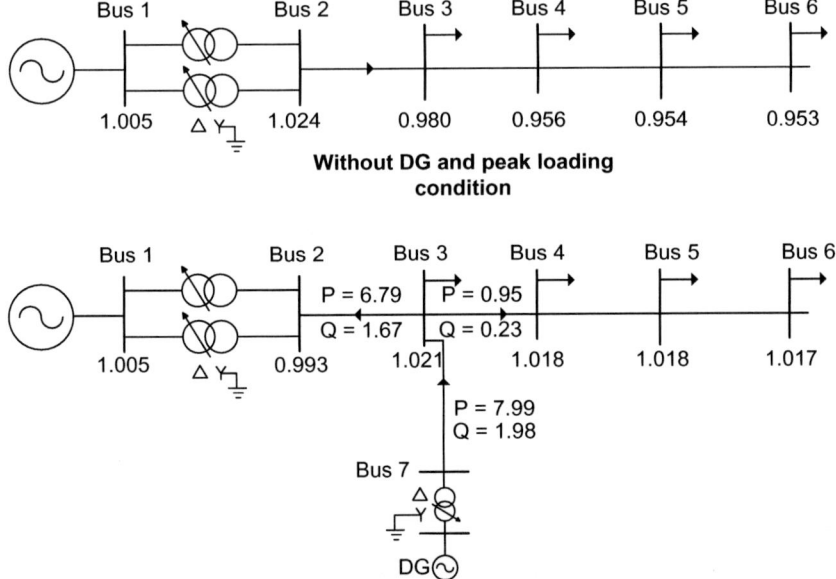

Figure 3.33   Voltages with LDC (P in MW and Q in MVAr)

Bus 6 becomes much higher than the reference setting. This can be explained using the voltage reference signal to the AVC that is given by Reference 16:

$$V_{ref} = V_M + I_P R + I_Q X \qquad (3.49)$$

where V_{ref} is the reference signal to the AVC, V_M is the measured voltage of the transformer secondary, I_P and I_Q are the active and reactive component of the current flowing from secondary side to primary side of the generator respectively, and R and X are the compensation resistance and reactance respectively.

When the DG is connected, both active and reactive currents of line section 2-3 reverse (see the lower diagram of Figure 3.33), whereas the power in other lines remains the same as in the previous case (but lower values due to minimum loading). The dominant effect comes from the line section 2-3 where I_P, I_Q, R and X are much greater than that of other line sections. The reverse power on this line section increases the reference setting of the AVC.

Thus, for this example, LDC continues to work satisfactorily with a distributed generator connected to one of the outgoing circuits. This may not always be the case with multiple outgoing circuits or if the sign of the $I_Q X$ term is made negative to give negative reactance compounding [16].

3.   Fault studies

A three-phase fault was applied to Bus 5 with and without DG and fault current flows in MVA are shown in Figure 3.34. It can be seen that with DG the fault current flowing through switchgear at busbars 3 and 4 has increased from 72 MVA to 91 MVA.

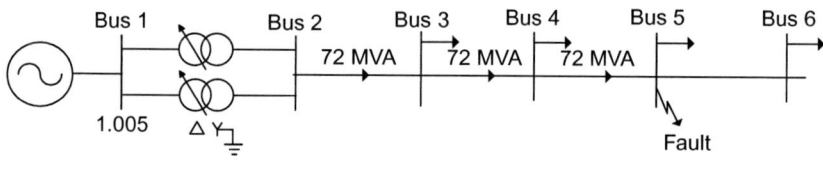

Without DG

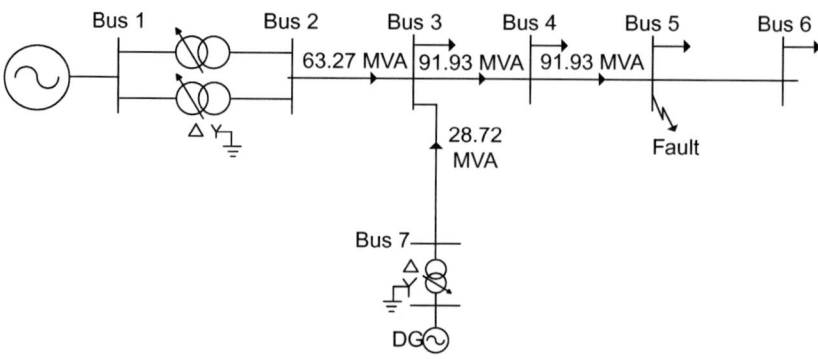

DG connected to Bus 3

Figure 3.34   Fault current flow to a three-phase fault at Bus 5

3.4.2 *Electromagnetic transient studies*

The use of mechanically switched capacitors for power factor correction of induction generators has a number of difficulties. If the degree of compensation approaches the no-load reactive power requirement of the generator, then there is a danger of self-excitation as the capacitors can only be switched occasionally and in discrete steps. The reactive power supplied by a fixed shunt capacitor reduces with the square of the network voltage so that during network voltage depressions, just when additional reactive power support is needed, the capability of fixed capacitors to provide reactive power is reduced. Conventional static var compensators (SVCs), which use thryristor switched capacitors, have a similar voltage/reactive power characteristic [17]. These limitations can be overcome if the reactive power required is supplied by a power electronic compensator rather than by capacitors. The STATCOM is a voltage source converter based reactive power compensator, which may be used to generate (or absorb) controlled reactive power [17]. It consists of a voltage source converter and a coupling reactor. The STATCOM operates by creating a voltage of controlled magnitude and phase at the converter terminals, and so exchanges real and reactive power with the network across the coupling reactor. Just as with a synchronous generator, the real power exchanged is controlled by the angle between the converter output voltage and the network voltage, while the reactive power exchanged is controlled by the relative magnitudes of the two voltages.

The VSC of the STATCOM is switched at a high frequency and the resulting output voltage is non-sinusoidal and contains harmonics. For accurate modelling of such system the load flow and fault calculation packages that solve a set of algebraic equations are not adequate. Usually a numerical integration of a set of differential equations is used to simulate this type of circuit. The analysis of this nature is called electromagnetic transient simulation and they normally demand a smaller time step for simulations.

Figure 3.35 shows the power circuit of the STATCOM. The voltage source inverter is a six pulse IGBT-based converter, which operates with space vector pulse width modulated (PWM) modulation. The STATCOM was simulated using the electromagnetic transient simulation programme EMTDC/PSCAD. Figure 3.36 shows the STATCOM terminal voltage (V_s), injected current (I_s) and current through a switch (I switch) when it is generating only reactive power.

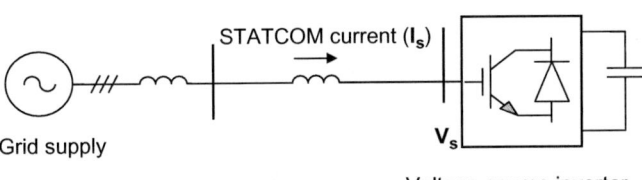

Figure 3.35   STATCOM used for simulations

STATCOM-ES

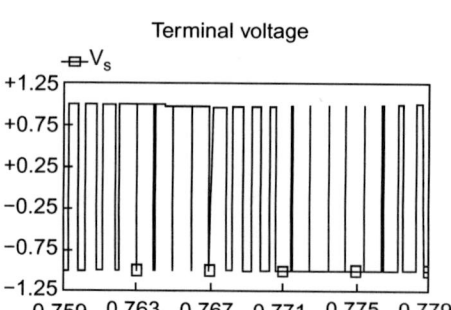

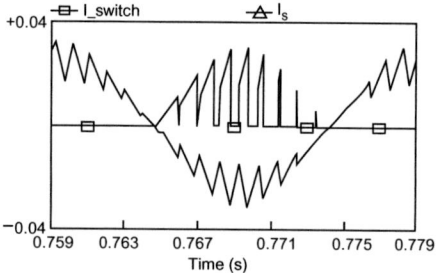

Figure 3.36   STATCOM voltage and current

A3.1   Appendix: Unbalanced faults

Line-to-line fault

For a line-to-line fault as shown in Figure 3.37, $I_A = 0$, $I_B = -I_C = I_f$ and $V_B - V_C = I_f Z_f$.

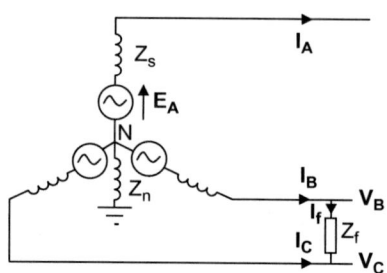

Figure 3.37   Line-to-line fault

From symmetrical components:

$$\begin{bmatrix} \mathbf{I_{A0}} \\ \mathbf{I_{A1}} \\ \mathbf{I_{A2}} \end{bmatrix} = \frac{1}{3} \begin{bmatrix} 1 & 1 & 1 \\ 1 & \lambda & \lambda^2 \\ 1 & \lambda^2 & \lambda \end{bmatrix} \begin{bmatrix} 0 \\ \mathbf{I_f} \\ -\mathbf{I_f} \end{bmatrix}$$

$$\therefore \mathbf{I_{A0}} = 0 \text{ and } \mathbf{I_{A1}} = -\mathbf{I_{A2}} \tag{3.50}$$

Substituting the symmetrical components of $\mathbf{V_B}$ and $\mathbf{V_C}$ into $\mathbf{V_B} - \mathbf{V_C} = \mathbf{I_f} Z_f$, the following equation was obtained:

$$\mathbf{V_{A1}}(\lambda^2 - \lambda) + \mathbf{V_{A2}}(\lambda - \lambda^2) = \mathbf{I_f} Z_f \tag{3.51}$$

Since $\mathbf{I_f} = \mathbf{I_B}$, from symmetrical components:

$$\mathbf{I_f} = \mathbf{I_{A0}} + \lambda \mathbf{I_{A1}} + \lambda^2 \mathbf{I_{A2}} \tag{3.52}$$

Substituting for sequence current from (3.50) into (3.52):

$$\mathbf{I_f} = (\lambda - \lambda^2)\mathbf{I_{A1}} \tag{3.53}$$

Substituting for $\mathbf{I_f}$ from (3.53) into (3.51) and divided by $(\lambda - \lambda^2)$, the following equation was obtained:

$$\mathbf{V_{A1}} - \mathbf{V_{A2}} = \mathbf{I_{A1}} Z_f \tag{3.54}$$

From (3.50) and (3.54), the connection of sequence networks for a line-to-line fault was obtained (Figure 3.38).

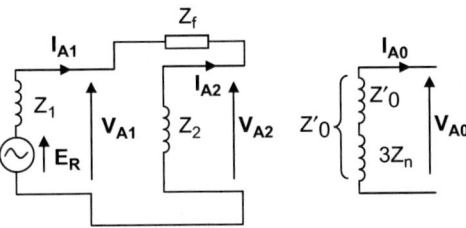

Figure 3.38   Connection of sequence networks under a line-to-line fault

Line-to-line-to-earth fault
For a line-to-line-to-earth fault shown in Figure 3.39, $\mathbf{V_B} = \mathbf{V_C} = 0$.

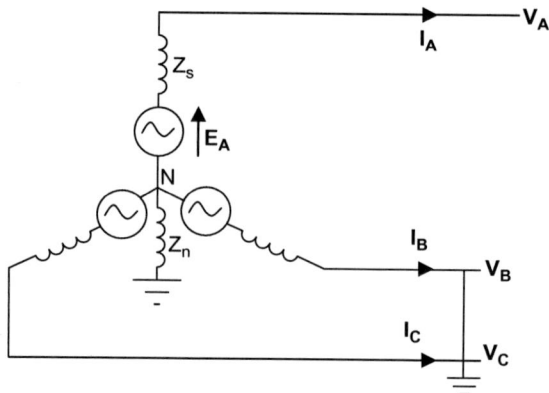

Figure 3.39   Line-to-line-earth fault

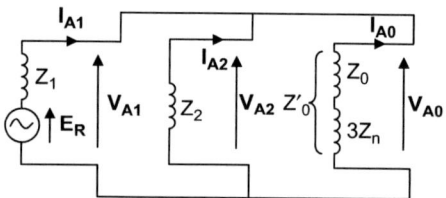

Figure 3.40   Connection of sequence networks under a line-to-line-earth fault

From symmetrical components:

$$\begin{bmatrix} \mathbf{V_{A0}} \\ \mathbf{V_{A1}} \\ \mathbf{V_{A2}} \end{bmatrix} = \frac{1}{3} \begin{bmatrix} 1 & 1 & 1 \\ 1 & \lambda & \lambda^2 \\ 1 & \lambda^2 & \lambda \end{bmatrix} \begin{bmatrix} \mathbf{V_A} \\ 0 \\ 0 \end{bmatrix}$$

$$\therefore \mathbf{V_{A0}} = \mathbf{V_{A1}} = \mathbf{V_{A2}} \tag{3.55}$$

Equation (3.55) suggests that three sequence networks should be connected in parallel as shown in Figure 3.40.

References

1. Chapmen S.J. *Electrical Machinery Fundamentals*. McGraw-Hill; 2005.
2. Hindmarsh J. *Electrical Machines and their Applications*. Pergamon Press; 1970.
3. McPherson G. *An Introduction to Electrical Machines and Transformers*. John Wiley and Sons; 1981.

4. Grainger J.J., Stevenson W.D. *Elements of Power Systems Analysis.* McGraw-Hill; 1994.
5. Weedy B., Cory B.J. *Electric Power Systems.* John Wiley and Sons; 2004.
6. Hurley J.D., Bize L.N., Mummert C.R. 'The adverse effects of excitation system Var and Power Factor controllers'. Paper No. PE-387-EC-1-12-1997. Presented at the IEEE Winter Power Meeting; Florida, 1997.
7. Heier S. *Grid Integration of Wind Energy Conversion Systems.* John Wiley and Sons; 1998.
8. Allan C.L.C. 'Water-turbine driven induction generators'. Proc. IEE, Paper No. 3140S, December 1959.
9. Olimpo Anaya-lara O., Jenkins N., Ekanayake J., Cartwright P., Hughes M. *Wind Energy Generation Modelling and Control.* John Wiley Press; 2009.
10. Burton T., Sharpe D., Jenkins N., Bossanyi E. *Wind Energy Handbook.* John Wiley and Sons; 2001.
11. Kundur P. *Power System Stability and Control.* McGraw-Hill; 1994.
12. Krause P.C., Wasynczuk O., Shudhoff S.D. *Analysis of Electric Machinery and Drive System.* John Wiley Press; 2002.
13. Ekanayake J.B., Holdsworth L., Jenkins N. 'Control of DFIG wind turbines'. *Power Engineer.* 2003;**17**(1):28–32.
14. National Grid Company plc. The Grid Code. Issue 3, Revision 25, 1 February 2008.
15. Anderson P.M. *Analysis of Faulted Power Systems.* IEEE Press; 1995.
16. Hingorani N.G., Gyugyi L. *Understanding FACTS: Concepts and Technology of Flexible AC Transmission Systems.* Wiley-IEEE Press; 1999.
17. Thomson M. 'Automatic voltage-control relays and embedded generation. Part II'. *Power Engineering Journal [see also Power Engineer].* 2000;**14**(3):93–99.

Chapter 4
Fault currents and electrical protection

Black Hill Wind Farm, Duns, Scotland (28.6 MW) [RES]

4.1 Introduction

Electrical faults, caused by the breakdown of insulation, are inevitable in any electrical power system. Faults may be caused by mechanical damage to the equipment or created by the degradation of insulation over time. Electrical protection is then used to operate circuit breakers that isolate the faulty equipment rapidly. Distributed generators must be protected against internal electrical faults, with fault current flowing from the network, and conversely the distribution network must be protected against fault current from the distributed generators. Islanded operation of smaller distributed generators is not generally permitted and so this condition is detected by protection systems designed to detect islanding or loss of mains and the generator then disconnected. Finally, the addition of distributed generation to a distribution network may alter the flows of network fault current in subtle ways and so lead to maloperation of conventionally designed distribution network protection systems.

When the insulation fails in a power system and a short-circuit fault occurs, excessive currents, perhaps as high as 20 times the load current, will flow.[1] These high currents may further damage the plant in which the insulation failure has occurred or damage the other items of equipment through which the fault current flows. These large fault currents can lead to fires or create hazardous voltages and are not allowed to persist for more than a second or two. Electrical protection schemes are used to isolate rapidly the faulty piece of plant while maintaining the sound pieces of equipment in service in order to ensure minimum disruption of supplies to customers. The large short-circuit currents disturb the operation of the power system, particularly by depressing the voltage, and so faults are isolated rapidly to maintain both voltage quality and stability of the system.

Different protection schemes have been developed to protect items of plant or sections of the distribution network. It is conventional to divide the distribution network into a number of zones to ensure discrimination is obtained and only the smallest possible section of network is isolated for any fault. Most distribution network protection is designed to respond to fault current that is supplied by the large central generators. These are electrically remote from faults on the distribution system and so the magnitude of fault currents is determined mainly by the impedance of the transmission and distribution networks. Thus, in traditional distribution network protection design, the fault currents are well defined and quite easy to calculate. The addition of generation located within the distribution network leads to more complex flows of fault current that no longer comes only from the transmission network. In addition many of the new types of generation use either induction generators or are connected to the network through power electronic converters, whose ability to provide fault current is determined both by the capability of the inverter and also by their control systems.

4.2   Fault current from distributed generators

All directly connected, spinning electrical machines (motors as well as generators) supply fault current into a short circuit in the network. Too much short-circuit current is dangerous, as it will over-stress circuit breakers that try to break the current and distort cables and other plant through which it passes. Too little short-circuit current from generators is also of concern, as most protection systems on distribution networks work by detecting the over-currents caused by faults, and so will not work correctly with too little fault current.

If a distribution system is designed to provide good power quality (i.e. limited variation in voltage with changes in load), it will have a high short-circuit level[2]

---

[1] The terms 'short circuit' and 'fault' are used interchangeably in this chapter.

[2] The short circuit or fault level is the short-circuit or fault current that would flow in the event of a short circuit or fault multiplied by the nominal prefault voltage at that point of the network.

that approaches the short-circuit rating of the switchgear and other components of the system. This is typically the case in the centres of cities and in industrial plants where the short-circuit level is kept high to minimise voltage changes caused by the high currents drawn when connecting large loads and starting large motors. If the short-circuit level is already high, then adding any distributed generation may lead to excessive short-circuit levels. In these circumstances the distributed generator will be refused permission to connect to the network.

The rapid and reliable detection of faults using short-circuit current from distributed generators is difficult due to their relatively small ratings and the long clearance times of distribution protection systems with only small fault currents. A small distributed generator, with the same per unit machine parameters on its rating as that of a large generator, will provide fault current only in proportion to the machine sizes and ratings. Distribution networks are often protected with time-delayed over-current protection that, because of the way it is set (or graded), can require fault currents considerably higher than the circuit continuous rating in order to operate quickly. Thus, the ability of a small generator to provide adequate fault current requires careful attention during the design of the distributed generation scheme.

4.2.1 Synchronous generators

The fault current of a synchronous generator into a three-phase fault depends on the rotor construction. If the machine is a salient-pole machine, both the direct axis (i.e. the axis of the field winding) and the quadrature axis reactances (synchronous, transient and sub-transient) contribute to the fault current [1,2]. For a cylindrical pole machine the direct axis has the same value of reactances as the quadrature axis, and the fault current contribution to a three-phase fault is usually described by an expression of the form:

$$I(t) = E_F \left[\frac{1}{X} + \left(\frac{1}{X'} - \frac{1}{X} \right) e^{-t/T'} + \left(\frac{1}{X''} - \frac{1}{X'} \right) e^{-t/T''} \right] \cos(\omega t + \lambda)$$

$$- \frac{E_F}{X''} e^{-t/T_a} \cos(\lambda) \tag{4.1}$$

where X = synchronous reactance, X' = transient reactance, X'' = sub-transient reactance, E_F = pre-fault internal voltage, T' = transient short-circuit time constant, T'' = sub-transient short-circuit time constant, T_a = armature (DC) time constant, λ = angle of the phase at time zero, ω = system angular velocity.

Note: Equation (4.1) is written in the conventional form [1,3–5, Tutorial II]. The sub-transient, transient and synchronous reactances are used to represent the performance of the machine at different times after the fault, defined by the corresponding time constants. The armature (DC) time constant is used to describe the decay of the DC component of the fault current.

The last term of (4.1) describes the DC component that depends on the point-on-wave at which the fault occurs, while the remainder describes the 50/60 Hz component. The short-circuit time constants (T' and T'') and the armature time constant (T_a) are not fixed values but depend on the location of the fault. In particular,

$$T_a = \frac{(X'' + X_e)}{\omega(R_a + R_e)}$$

(4.2)

where X_e = external reactance (to the fault), R_e = external resistance (to the fault), R_a = armature resistance.

Synchronous machine impedances have X/R ratios that are much larger than those of distribution circuits. Hence, a fault close to a synchronous generator will have an armature time constant (T_a) and DC component that lasts much longer than for a remote fault. This is an important consideration for distributed generation schemes. Traditional, passive distribution systems, fed from high-voltage (HV) networks through a succession of transformers and circuits with some resistance can be considered to have fault currents with a very rapid decay of the DC offset and an essentially constant AC component. In contrast, faults close to generators, or large motors, will have slower decaying DC offset and decaying AC components. This is recognised in IEC or BS EN 60909 [6] that recommends two different calculation approaches for these two situations, i.e. the 'far-from-generator-short-circuit' and the 'near-to-generator-short-circuit'. Engineering Recommendation G74 [7] also discusses the various computer-based modelling approaches that may be used to represent this effect.

Figure 4.1 shows a simulation of the response of a synchronous generator to a close-up and remote fault. (For clarity, only the phase with maximum offset is shown.) Obviously the fault current is much larger for the close-up fault than for the remote fault. The close-up fault shows the longer decay in the DC component and the reducing AC component. The remote fault shows the very rapid decay of the DC component and only a very small reduction in the AC current with time.

A circuit breaker close to a synchronous generator has a more onerous duty when breaking fault current than one in a passive distribution network and so old distribution circuit breakers, which were never intended to be used close to synchronous generators, may need to be changed when distributed generation is installed.

Figure 4.2 shows a curve that is a typical of those supplied by manufacturers to describe the fault current capability of a small synchronous generator on to a three-phase fault at its terminals. In this case the current is expressed in rms values with logarithmic scales on the axes. It may be seen that the expected decay occurs up to, say, 200 ms but then the excitation system operates to boost the fault current back up to three times full-load output (by increasing E_F in (4.1)). This is necessary as distribution protection is usually time-graded and so sustained fault current is required if current operated protection is to function effectively. This ability to boost the fault current depends critically on the excitation scheme that has been chosen. Depending on the generator design, a 3 per unit (pu) sustained fault current on to a terminals short circuit may require 'field forcing' to an internal voltage of 8–10 times that needed at no-load.

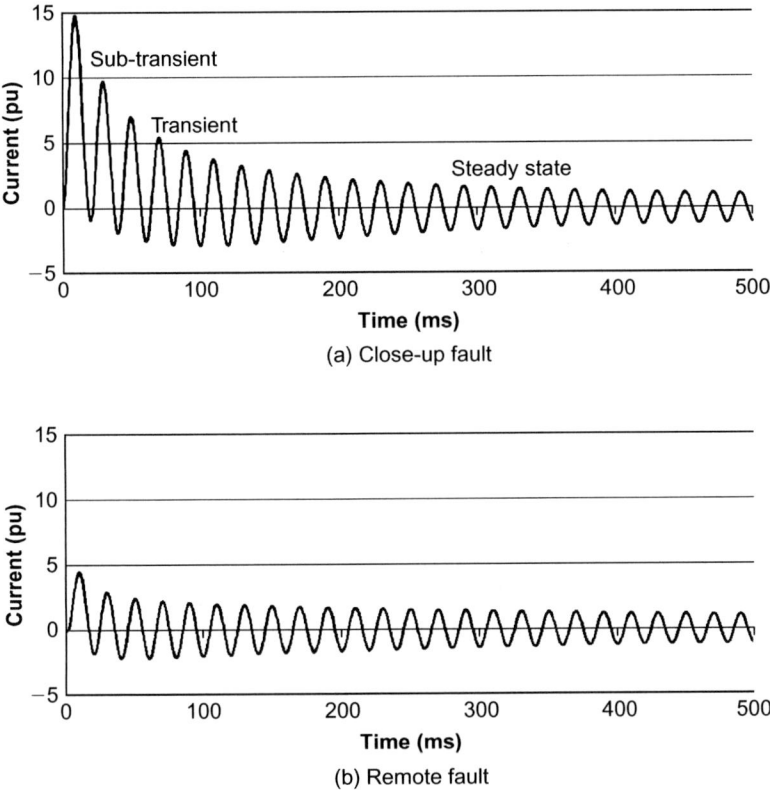

Figure 4.1   *Fault current of a synchronous generator (phase with maximum offset)*

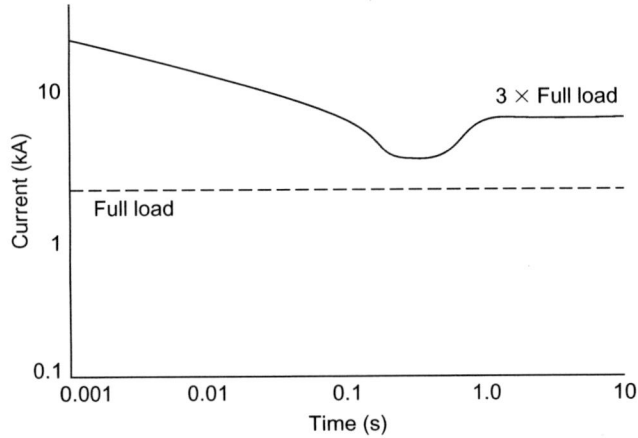

Figure 4.2   *Short-circuit decrement curve for a small synchronous generator*

A paper by Griffith [8] gives an excellent review of various types of generator excitation including a discussion of field forcing and the possible use of voltage-controlled over-current protection for small distributed generator schemes.

4.2.2   *Induction generators*

The behaviour of an induction generator under fault conditions is rather different to that of a synchronous generator. The operation of an induction machine with a three-phase fault on the network is described using the transformer equivalent circuit of an induction machine shown in Figure 4.3 (see also Figure II.14). In Figure 4.3 the rotor resistances and reactances are locked rotor values. In this book, it has been assumed that these values are constant and do not depend on the slip.[3] E_2^r refers to the induced emf when the rotor is rotating and s is the slip.

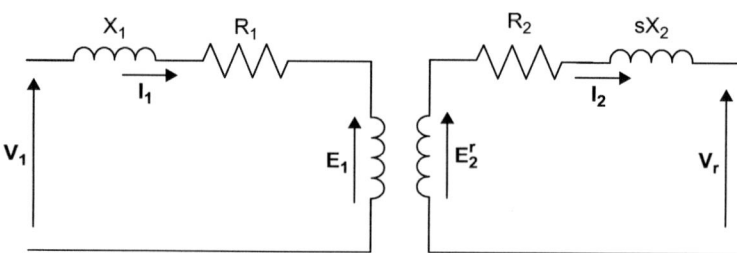

Figure 4.3   Induction machine equivalent circuit (motor convention). The rotor is short circuited in a squirrel-cage machine, V_r is then zero.

If a fault occurs on the power system, the terminal voltage, V_1, will be reduced (the magnitude of the reduction depends on the location of the fault). However, immediately after the fault, the magnetic field, ϕ, will not change, thus E_1 will be same as its pre-fault value. So the stator current will be reversed and the induction machine feeds into the fault. The magnitude of fault current will depend on the relative magnitudes and phases of E_1 and V_1.

Under normal operating conditions or for a remote fault (where V_1 is sufficiently large to supply the magnetising current), the stator produces a magnetic field which rotates at ω_s. When the rotor is rotating at ω_r, there will be relative motion between the rotor conductors and the stator magnetic field equal to $\omega_s - \omega_r = s\omega_s$. Thus the magnitude of the rotor-induced emf E_2^r is given by:

$$E_2^r = s\omega_s\phi k \tag{4.3}$$

where ϕ is the magnetic field at the air gap and k is a proportional constant.

---

[3] For a single-cage rotor construction, the rotor resistance and reactance are essentially constant. However for a double-cage (deep bar) rotor construction, the rotor resistance decreases with the slip, and the rotor reactance increases with the slip [9]. During normal operation the slip (s) is small (around 1%) and the frequency of the induced rotor currents is low. Hence, the currents are distributed across the depth of the rotor bars to give a low value of resistance and a high value of reactance. When the slip approaches 1 (locked rotor conditions) the currents are concentrated near the surface of the rotor by the skin effect, and the rotor resistance increases and the rotor reactance decreases.

The rotor current is given by:

$$I_2 = \frac{E_2^r}{R_2 + jsX_2} \quad (4.4)$$

The situation is different for a three-phase fault at the terminals of the induction generator. Due to the fault, V_1 becomes zero. The magnetic field (ϕ) will not diminish immediately after the fault but will cease to rotate. As the rotor continues to rotate at ω_r, the relative movement between the rotor conductors and the magnetic field is equal to ω_r. Thus the magnitude of the rotor-induced emf $E_{2\_fault}^r$ is given by [9,11]:

$$E_{2\_fault}^r = \omega_r \phi k \quad (4.5)$$

Then, the frequency of the rotor-induced emf is equal to $\omega_r/2\pi$. Therefore, the rotor reactance will be $(\omega_r/\omega_s)X_2$ (as the locked rotor value X_2 is calculated for the synchronous frequency, $\omega_s/2\pi$). Thus, the rotor current during a fault is given by:

$$I_{2\_fault} = \frac{E_{2\_fault}^r}{R_2 + j(\omega_r/\omega_s)X_2} \quad (4.6)$$

From (4.3) to (4.6) and with $\omega_r/\omega_s \approx 1$:

$$\frac{I_{2\_fault}}{I_2} \approx \frac{1/s}{1 + j(X_2/R_2)} \quad (4.7)$$

Using (4.7), for a 2 MW induction generator with $R_2 = 0.0055$ pu and $X_2 = 0.1$ pu, it was calculated that the fault current (immediately after the fault) at an operating slip of 1% is about 5.5 times the rated current. However, as there is no reactive power source to sustain the magnetising current, the magnetic field collapses (thus E_2^r collapses) and the fault current decays to zero. Figure 4.4 shows one phase of a simulation of a 2 MW, 690 V induction generator. It may be seen that the fault current decays within 100–200 ms.

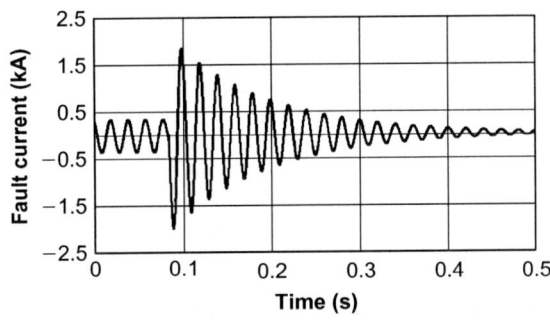

Figure 4.4   Fault current of an induction generator with a three-phase fault at its terminals (phase shown with minimum DC offset)

An expression similar to (4.1) may be used to describe the fault current contribution of an induction generator to a three-phase fault at its terminals. However, due to difficulties with obtaining data it is often reduced to:

$$I(t) = \frac{V_1}{X''} [\cos(\omega t + \lambda)e^{-t/T''} + \cos(\lambda)e^{-t/T_a}] \tag{4.8}$$

where (see Figure II.15 for resistances and reactances)

$$X'' = X_1 + \frac{X_2'X_m}{X_2' + X_m}$$

$$T'' = \frac{X''}{\omega R_2'}$$

$$T_a = \frac{X''}{\omega R_1}$$

V_1 is the magnitude of the network voltage (sometimes increased by a safety factor to account for the voltage variations depending on the time and place, changing of transformer taps, neglecting load and capacitance, and the sub-transient behaviour of generators and motors [7]).

$\omega = (1 - s)\omega_s \approx \omega_s$ (as s is very small)

X_2' and R_2' are the stator-referred rotor reactance and resistance.

As for a synchronous generator, any external impedance involved must be added to the stator impedance.

Unbalanced faults on the network may lead to sustained fault currents from induction generators and in some cases a rise in current on the unfaulted phase(s). Appropriate computer simulations are required for an accurate representation of the behaviour of an induction generator feeding a sustained unbalanced fault.

The fault current from an induction generator is generally not relied upon for the operation of any protective relays. Therefore, when a fault occurs on a distribution system connected to an induction generator, the fault current from the network source is used to operate the distribution system over-current protection. This isolates the generator and so over-voltage, over-frequency or loss-of-mains relays are then used to trip its local circuit breaker and prime mover. This sequential tripping of the generator using voltage, frequency or over-speed protection is necessary as the induction generator is not capable of providing reliable, sustained fault current.

4.2.3   Doubly fed induction generators

The fault current contribution of DFIGs is discussed in References 9, 11 and 12. Their fault current contribution depends significantly on the pre-fault operating condition. Assuming that the rotor-injected voltage is unchanged immediately after a fault, the DFIG fault current is explained using Figure 4.3 (in this case the rotor is not short circuited and the converter connected to the rotor continues to inject a voltage). When a three-phase short circuit occurs at the terminals of the DFIG, the induced emf on the rotor circuit is given by (4.5). The subsequent rotor fault current will be governed by the difference between the rotor-induced emf and the injected voltage.

If the back-to-back power electronic converters connected to the DFIG rotor are rated to take large fault currents, then there exist three current components on the stator [9,11].

1. A DC component on the rotor circuit decaying at a time constant of $T'' = X_2''/\omega R_2'$ (where $X_2'' = X_2 + (X_1 X_m/X_1 + X_m)$ and R_2' is the stator-referred rotor resistance). The magnetic field created by this current rotates with the rotor and creates an alternating current in the stator.
2. A DC component on the stator circuit that decays at a time constant of $T_a = X_1''/\omega R_1$ (where $X_1'' = X_1 + (X_2 X_m/X_2 + X_m)$).
3. A sustained fault current due to the voltage fed by the rotor converters (the term $I_m \cos(\omega t + \lambda)$ in (4.9)).

Therefore, an expression similar to (4.8) may be obtained for the DFIG with sufficiently highly rated converters [9,11]:

$$I(t) = \frac{V_1}{X''}\left[\frac{X_m^2}{X_1 X_2}\cos(\omega t + \lambda)e^{-t/T''} + \cos(\lambda)e^{-t/T_a}\right] + I_m \cos(\omega t + \lambda) \quad (4.9)$$

In this equation, $\omega = (1 - s)\omega_s$ and s depends on the pre-fault operating condition. The value of s may vary between $+0.4$ and -0.2 (this depends on the wind turbine design).

Figure 4.5 shows a simulation of the stator and rotor current of a DFIG when the generator is operating with a constant mechanical torque of $T_m = 1.0$ pu and a three-phase fault was applied at $t = 10$ s.

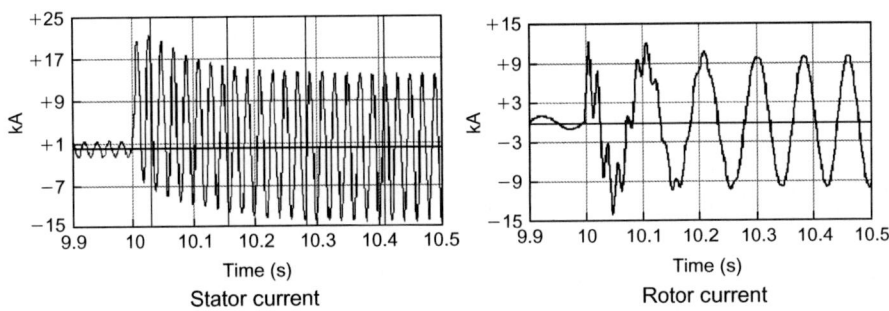

Figure 4.5 Simulation of fault current of a DFIG without crowbar operation [13]

Practical DFIG converters are often designed to carry only a limited over-load current. Therefore, excessively large rotor currents are not allowed to flow in the rotor circuit. In order to protect the power electronic converters, a crowbar is often employed that short circuits the rotor circuit through an impedance. The subsequent operation of the DFIG is then very similar to an induction generator with increased rotor impedance (rotor impedance plus crowbar impedance). Therefore, (4.8) can be used for the DFIG fault current with increased R_2' and/or X_2' (this depends on the crowbar impedance, but typically a resistor is used in the crowbar circuit) and $\omega = (1 - s)\omega_s$.

4.2.4   *Generators connected through power electronic converters*

IEC/BS EN 60909 [7] requires that power electronic motor drives that can operate regeneratively are considered for their contribution to initial short-circuit current but not to any sustained short-circuit current. They are represented as an equivalent motor with a locked rotor current of $3\times$ rated current. The CIGRE report on dispersed generation [14] indicates that a fault current only equal to rated current may be expected.

The actual behaviour of the generator/converter system when supplying fault current depends both on its power circuit and control system. To protect the semiconductor devices in a converter against large transient current flowing during a fault, an instantaneous over-current shut-down protection element is normally employed [15]. When the fault current reaches the over-current trip level, the over-current protection activates within a few microseconds. Therefore, the fault current contribution from a converter-connected distributed generator is very low. If the power circuit of the converter consists of thyristors (current source converter), then the fault current contribution during the first few cycles may be about 2–3 times the rated current of the converter. If the converter is IGBT based (voltage source converter), then the converter will not generally be able to provide a fault current of more than 120% higher than its rating. However, there are voltage source converter-based systems that can provide up to 2–3 times the rated current for a short time.

For a voltage source converter to continue to operate under unbalanced faults the control systems (including the phase locked loop) must be robust. Simple designs of voltage source converters used with small distributed generators are unlikely to stay operating during network unbalances.

4.3   Fault current limiters

Fault currents of the network, after the connection of a new distributed generator, may be greater than the short-circuit capability of the existing switchgear and cables. This may require replacing the switchgear (and other equipment) with that of a higher short-circuit rating. However, such upgrades are not easy due to: (a) high replacement cost, (b) disruption to the supply during the construction period and (c) the short-circuit ratings of presently available plant may soon become inadequate. Other than replacing the existing switchgear, there are several methods that can be used to reduce the fault current in the power system to an acceptable level. These methods include: (a) network splitting, (b) current limiting reactors, (c) current limiting fuses and (d) fault current limiters [16,17].

4.3.1   Network splitting

If the network is split, often by opening substation bus-section circuit breakers, the sources of short-circuit current are separated thus increasing the impedance between the sources and the fault. This in turn reduces the fault current. The consequence of network splitting is that the number of duplicate supply paths to a load is reduced with potential reduction of reliability [17].

4.3.2   Current limiting reactors

A reactor can be placed in series with the generator circuit to increase the impedance to the fault and so reduce the fault current. Despite a voltage drop and a power loss associated with a reactor during normal operation, current limiting reactors are used in practice due to their relatively low cost of implementation and maintenance.

4.3.3   Current limiting fuses

Current limiting fusegear consists of two parallel conductors: a main conductor and a parallel fuse. Under normal operation, the load current flows through the main conductor. During a fault, a tripping device disconnects the main conductor, transferring the fault current to the parallel fuse with a high breaking capacity, which limits the fault current during the initial rise of the first 50/60 Hz cycle [2].

4.3.4   Fault current limiters

Fault current limiters (FCLs) have a high impedance during a fault while having a low impedance under normal operating conditions. Depending on the technology used to obtain the required impedance characteristic, FCLs can be categorised into three main groups: (a) superconducting fault current limiters (SFCL) [16,18,19], (b) magnetic fault current limiters (MFCL) [20,21] and (c) static fault current limiters [22,23].

There are two major types of the SFCL namely, resistive and inductive. The resistive type is essentially a superconductor connected in series with the power circuit [18,24,25]. The superconductor shows a negligible resistivity below a critical temperature (T_c) and a critical current density (J_c) (superconducting state), and as soon as T_c or J_c are surpassed, the resistivity of the material increases rapidly (normal state) [26]. The superconductor is located in a liquid nitrogen bath referred to as 'cryostat'. At liquid nitrogen temperature (77 K), the superconductor shows a negligible resistance. During a fault as a high current passes through the superconductor, both current density and temperature increase above T_c and J_c. This in turn takes the superconductor to its normal state, thus showing a high resistance in series with the power line. This series resistor effectively limits the fault current.

Inductive SFCLs can broadly be divided into screened-core and saturated-core types [18,25]. The screened-core design has a primary copper winding connected in series with the line, a superconducting cylinder (which forms a single turn second-ary) and an iron core. The primary winding in this device is usually wound over the superconducting cylinder. Therefore, the superconductor acts as a magnetic shield or screen preventing primary winding flux from entering the iron. This can be visualised as a transformer with a short-circuited secondary thus giving negligible impedance in series with the line [18]. During a fault the superconductor moves to its normal state, thus the transformer primary reactance acts in series with the line.

A saturated-core SFCL and some MFCL designs operate essentially on the same basic principle. A simple construction of a saturated-core FCL is shown in Figure 4.6.

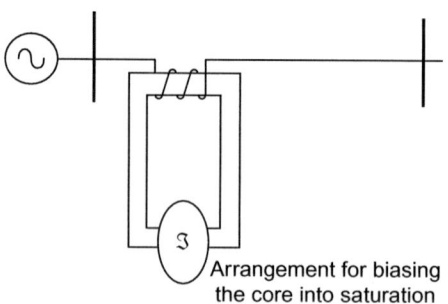

Arrangement for biasing
the core into saturation

Figure 4.6   Saturated-magnet SFCL and its operating principle

The magnetic core of the winding is biased into saturation by a DC magnetic field produced by a superconductor [25] or by a permanent magnet [21]. With low operating currents, the device presents only a low saturated inductance to the cir-cuit. During a fault the large fault current brings the core out of saturation, thereby presenting a high impedance to the power circuit. A single core only acts for one half of the AC waveform, and a similar device that is acting on the other half cycle of the AC waveform is required to provide current limiting in both cycles [21].

A static fault current limiter is shown in Figure 4.7. In normal operation the thyristor switch is OFF and current flows through Reactor 1 and the capacitor. The reactances of the capacitor and Reactor 1 are chosen to be equal so that the series combination shows very small impedance. During a fault the thyristor switch is turned ON, the capacitor is effectively by-passed and the impedance of the circuit becomes reactive thus limiting the fault current.

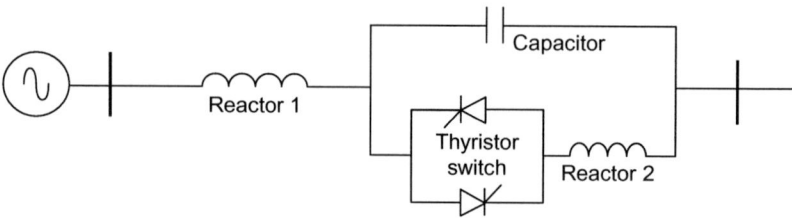

Figure 4.7   A static fault current limiter

4.4 Protection of distributed generation

Traditionally, power-system protection has been designed assuming that central generation feeds the distribution network and thus fault current always flows from the higher to the lower voltage levels. However, with the introduction of distributed generation, both central and distributed generators feed current into a fault. This multi-directional flow of fault currents requires the rechecking of existing protection coordination and reach.

A fault on the distribution system may result in a distributed generator being disconnected together with some loads, thus creating a power island. As the fault current from a distributed generator can be very low, a subsequent fault on the islanded system may not be detected. In addition, depending on the design of the network, the connection of the system neutral to earth (ground) may be lost during islanding. Both conditions are undesirable. Additionally the creation of power islands leads to difficulties with the use of auto-reclose on distribution networks as well as posing safety issues for maintenance staff. Therefore, the protection philosophy of a distributed generator should determine when it should stay connected, supporting the main power system, and when it should be tripped off to ensure safety.

There is a well-known inconsistency in the way smaller generators are considered for connection to distribution and transmission networks, which will become more and more important as the deployment of distributed generation increases. Smaller generators that are connected to distribution networks are governed by the requirements of IEEE 1547 (in the United States) [27] and G59 (in the United Kingdom) [28]. IEEE 1547-2003 is applicable to all single or multiple distributed generators of aggregated capacity of 10 MVA or less, and G59/1 applies to generators of 5 MW or less which are connected at 20 kV or below. These standards require distributed generators to trip off during network faults/disturbances and not to support power islands. In contrast for larger generators, G75 [29] allows islanded operation to be considered, and the transmission Grid Codes require that non-conventional generation (e.g. larger wind farms) connected to transmission systems must remain stable and connected during network faults and support the operation of the wider power system. As more and more small distributed generation is connected to the power system, its aggregate effect will become important and the consequences of it tripping during a network disturbance unacceptable. There have already been instances reported in mainland Europe and Great Britain of a system disturbance caused by network faults and loss of central generation being made worse by the subsequent disconnection of distributed generation.

Figure 4.8 shows a distribution circuit with a distributed generator and typical earthing arrangements used in the United Kingdom [30]. The protection philosophy associated with the generator connection is explained by considering faults in different parts of the network.

1. For a phase or earth fault inside the distributed generator, fault currents flow from the distribution network and that current is used to detect the fault.
2. For a fault at F_1 (i.e. on the connection between DG and the distribution network), circuit breaker B removes the fault current flowing from the distribution

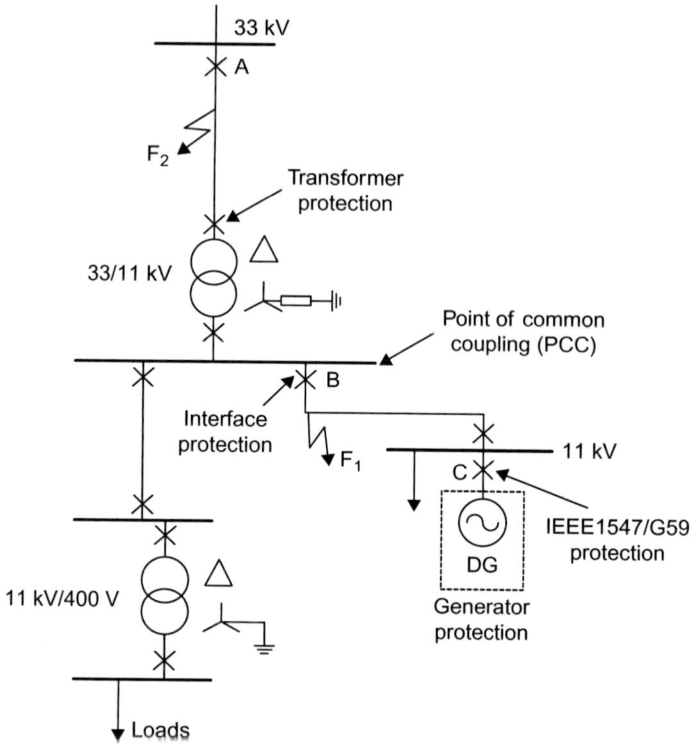

Figure 4.8   Typical connection of a distributed generator up to 33 kV

network. The over-current protection of circuit breaker C will attempt to detect the fault current from the generator and disconnect the generator. However, detecting the fault current from the distributed generator depends on the type of the machine used. If DG is a synchronous generator, the fault current may be adequate to detect the fault using an over-current relay. However, if DG is an induction generator or a converter-connected generator, its fault current is not adequate to detect the fault by an over-current relay. Then, other methods must be used to detect and isolate the fault.

Often when there is a fault at F_1, the voltage at the generator terminals will be suppressed and an under-voltage relay may be used to detect the fault. However, if the interconnector is long, the reduction in voltage may not be adequate to detect the fault by the under-voltage relay. A fault at F_1 prevents power being fed into the distribution network from the generator, thus depending on the adjacent loads, the frequency of the generator network (the generator and any captive loads) may increase or decrease. An under or over-frequency relay is then used to detect this condition and trip the generator.

3. Disconnection of a part of the network with DG due to a fault on the distribution network has the potential to lead to isolated operation of the distributed generator or islanding. For example, a fault at F_2 will cause circuit breaker A to trip and thus

the distributed generator and part of the network may then operate as an island. This condition is generally unacceptable, and protection based on rate of change of frequency (ROCOF) is used to detect this loss of mains and trip the generator.

4.4.1 *Protection of the generation equipment from internal faults*

A generator may malfunction due to insulation failure in the stator or rotor winding, due to a fault in the prime mover, excitation system (of synchronous generators) or due to a mechanical failure such as a loss of cooling or bearing failure.

4.4.1.1 Protection of the generator stator

The stator of a small generator is usually provided with over-current and thermal protection. An over-current may be caused by an earth fault, an inter-turn fault or a phase fault. A short circuit between the winding and core leads to an earth-fault, and is a common fault. An inter-turn fault is rare occurring inside the stator winding where one or more turns of the winding are close together. In small generators, an inter-turn fault may only be able to be detected if it develops into an earth fault. If each slot of a generator stator carries more than one winding, then there may be the possibility of a phase-to-phase fault. Another possibility is a short circuit in the end windings.

Earth-fault protection of a generator depends on the grounding method used. For a star-connected generator, if the star point is directly earthed or earthed via a low impedance, the earth-fault current is approximately equal to the phase-fault current. However, as the earthing impedance increases, the earth-fault current may not be large enough to be detected by an over-current relay.

For a generator with low impedance earth, the simplest form of earth-fault protection may be provided by connecting an earth-fault over-current relay to the neutral connection as shown in Figure 4.9. The earth-fault relay is not affected by load current and only sees the residual current of an earth fault. Therefore, the relay is set to operate at 20% of full-load current of the generator.

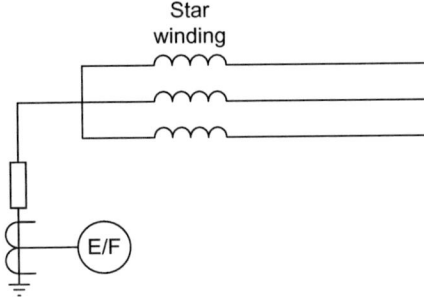

Figure 4.9   Earth-fault protection of a generator stator

As shown in Figure 4.10, a differential relay may be employed to detect winding to earth faults, IEEE Relay Type 87 [31]. The differential relay compares the neutral current $(\mathbf{I_N})$ with the addition of three-phase currents $(\mathbf{I_A} + \mathbf{I_B} + \mathbf{I_C})$. Under normal operating conditions or for faults outside the protected zone (the protected zone is between the neutral current transformer (CT) and phase CTs), $I_A + I_B + I_C = I_N$. Therefore, the current through the differential relay is zero. If there is an earth fault within the protected zone, $I_A + I_B + I_C \neq I_N$ and a current flows through Relay 87, thus tripping the generator.

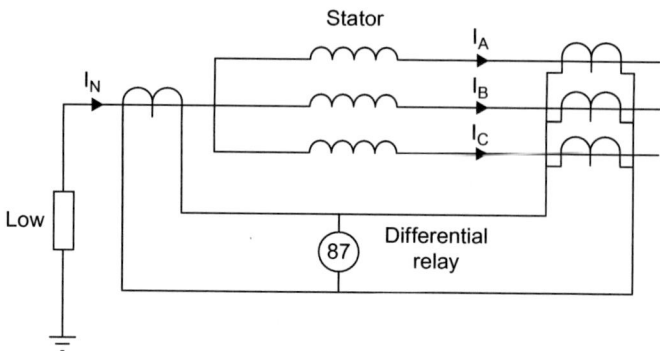

Figure 4.10   Earth-fault protection of a generator

The minimum pickup setting of the differential relay should be adjusted to sense faults on as much of the winding as possible. However, settings below 10% of full-load current carry increased risk of maloperation due to transient CT saturation during external faults or during step-up transformer energisation.

The ground differential (shown in Figure 4.10) works well for low impedance grounded generators. However, the portion of the winding protected by this method may not be adequate for high impedance grounding.

EXAMPLE 4.1

The star point of a 6.6 kV three-phase alternator is earthed through a non-inductive resistance of 1.75 Ω. The reactance of the alternator is $j5$ Ω/phase, and its resistance is negligible. As shown in Figure E4.1 circulating current protection is used and the relay is set to operate at 0.5 A. Determine the CT ratio, which will ensure 90% of the winding is protected. With the CT ratio calculated, what percentage of the winding is protected if the value of the earthing resistor is increased to 3 Ω? Assume that the generator is connected through a delta–star transformer to the network (delta on the generator side).

Answer:

The delta–star transformer blocks zero sequence current from flowing into the fault from the distribution network, thus fault current circulates between the generator earth and the faulted section of the winding (see Figure E4.1).

Since 90% of the winding is protected, the fault current for 10% of the stator winding is less than the current required for relay operation.

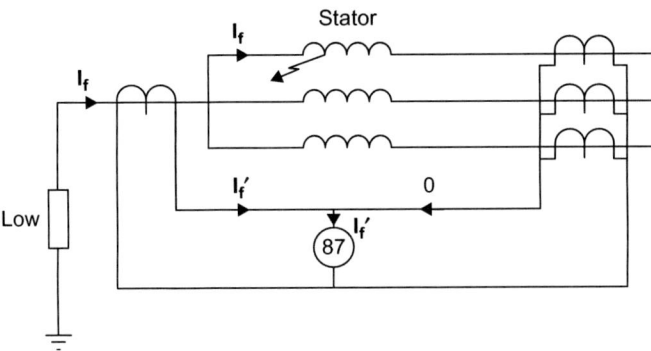

Figure E4.1

Assuming that the reactance of the alternator and its internal voltage are 10% of their fully rated values:

$$\text{Phase voltage of the 10\% of the winding} = \frac{10}{100} \times \frac{6.6 \times 10^3}{\sqrt{3}} = 381.05\,\text{V}$$

$$\text{Reactance of the 10\% of the winding} = \frac{10}{100} \times 5 = 0.5\,\Omega$$

From I = V/Z, the fault current for 10% of the winding is given by:

$$I_f = \frac{381.05}{[1.75^2 + 0.5^2]^{1/2}} = 209.4\,\text{A}$$

The relay operates when the neutral current exceeds 0.5 A. Therefore for 90% of the winding to be protected

0.5 < 209.4/CT ratio

CT ratio < 418.73

Therefore a CT ratio of 400/1 was selected.

When the value of the earthing resistor is increased to 3 Ω, consider that x% of the winding is protected. Then the fault current is greater than the current required for relay operation for $(100 - x)$% or higher of the stator winding. Similarly to the previous case:

$$\text{Phase voltage of the}\,(100 - x)\%\,\text{of the winding} = \frac{(100 - x)}{100} \times \frac{6.6 \times 10^3}{\sqrt{3}}\,\text{V}$$

$$\text{Reactance of the}\,(100 - x)\%\,\text{of the winding} = \frac{(100 - x)}{100} \times 5$$

For the relay to operate:

$$\frac{((100-x)/100) \times ((6.6 \times 10^3)/\sqrt{3})}{[3^2 + (((100-x)/100) \times 5)^2]^{1/2}} > 0.5 \times 400\,A$$

$$x < 83.7$$

That is when the earthing resistance is increased to 3 Ω, only 83.7% of the winding is protected.

Figure 4.11 shows a generator with high impedance grounding and a delta-connected step-up transformer. The high impedance ground connection is made using a distribution transformer so that the value of the loading resistor referred to the primary limits the earth-fault current. This arrangement allows the use of a more robust, lower value resistance [32]. For phase-ground faults on the generator windings, the current is not adequate to operate a differential relay. Hence, a voltage relay is used across the secondary of the distribution transformer. The detection of faults near the generator neutral requires a low over-voltage relay setting. However, the voltages associated with the low earth-fault currents may be comparable to voltages caused by 3rd harmonic currents in the neutral and so the relay is made insensitive to 3rd harmonic voltages.

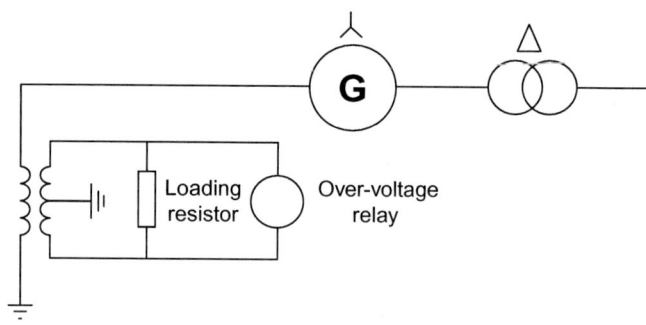

Figure 4.11   Earth-fault protection of high impedance grounded generator

Combined phase and earth-fault protection for a generator stator can be achieved by differential protection [32]. A high impedance relay or a relay with a biasing coil can be applied for differential protection as shown in Figure 4.12. In both cases, incoming current is compared with the outgoing current. If there is a phase-to-phase fault or a phase-to-earth fault within the protected zone, there will be a difference in the two currents and that is used to activate the relay. The high impedance relay is normally provided with a stabilising resistor to prevent nuisance tripping.

The stators of distributed generators are often provided with thermal protection. The winding temperature can increase beyond a preset value, normally around 120 °C, due to a sustained over-load or due to a failure of the cooling system.

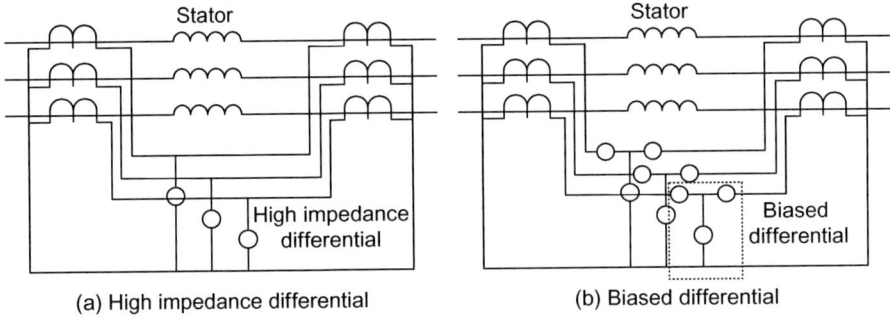

Figure 4.12   Differential protection of generators

Thermal sensors embedded in the stator windings are then used to activate an alarm or trip the machine. The difference of the input and output temperature of the cooling fluid may also be used to detect overheating. For small generators, a 'replica' type thermal estimation device may be provided to estimate the actual machine temperature using load current measurements.

4.4.1.2   Protection of generator rotor

The rotor of a synchronous generator carries the ungrounded DC field winding, and any single fault between the rotor conductors and rotor core can remain undetected and the machine continue to operate normally. However, a second earth fault will short-circuit part of the rotor winding, thus distorting the field across the air gap. This may create unbalanced forces on the rotor, vibration or thermal damage to the field winding. The best way to prevent dual earth faults on a generator rotor is to detect the first earth fault and trip the generator field and main circuit breaker.

Different methods are used to detect an earth fault on the rotor. A commonly employed method is by injecting AC or DC to the rotor as shown in Figure 4.13.

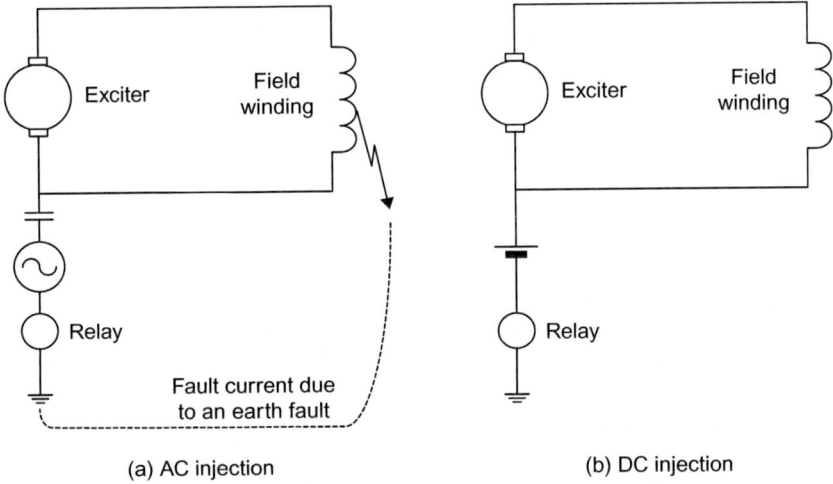

Figure 4.13   Rotor injection method for detecting earth faults

A DC source is directly connected to the rotor, whereas an AC source is connected via a capacitor. Whenever there is an earth fault on the field winding, the circuit with the current injecting source will be completed and a current flows through a sensitive current relay. In some generators, a potentiometric method that measures the voltage to the earth at the midpoint of a centre tap resistor is employed to detect earth faults in the field winding. This method works except for an earth fault very closer to the midpoint of the field winding.

4.4.1.3   Loss of excitation protection

A loss of excitation causes the rotor magnetic field to collapse. If damper windings are present on the rotor, the generator continues to operate as an induction generator, where short-circuited damper windings create the rotor magnetic field in the same way as an induction machine. It is estimated that a small generator of 50 MVA or less can operate as an induction generator safely for 3–5 minutes [33]. Depending on the generator loading, a large slip frequency current may flow in the rotor circuits, thus causing overheating. Further to deliver the power demanded by the governor (set to provide the initial loading), the slip of the generator may increase, thus leading to loss of synchronism. One of the other drawbacks of induction generator operation is a large amount of reactive power it draws to set up the magnetic field. On a weak network, this may depress the voltage of the nearby busbars or may lead to voltage collapse.

Protection against loss of excitation is normally provided by an impedance or admittance relay. Consider a case where the generator supplies apparent power of $S = P + jQ$ at its terminal voltage $\mathbf{V}$. The current through the stator windings is given by $\mathbf{I} = (P - jQ)/\mathbf{V}^*$. If an impedance relay is connected at the terminal of the generator, then the impedance seen by the relay is given by:

$$\mathbf{Z} = \frac{\mathbf{V}}{\mathbf{I}} = \frac{\mathbf{V}\mathbf{V}^*}{P - jQ} = \frac{V^2(P + jQ)}{P^2 + Q^2} = \frac{V^2(P + jQ)}{S^2}$$

$$= \frac{V^2}{S}(\cos\phi + j\sin\phi)$$

Under normal operation, $\cos\phi$ varies in between 0.9 lagging and 0.9 leading (this depends on the capability of the generator), thus $\sin\phi$ is small. However, when the excitation is lost as the machine draws a large amount of reactive power, $\sin\phi$ increases, thus changing the impedance seen by the relay. This change in impedance is used to detect the loss of excitation.

4.4.1.4   Loss of prime mover

When the prime mover fails, a generator can continue to operate as a motor. In this condition, a steam turbine might over-heat due to the reduction of steam flow across the turbine blades or loss of cooling system. In the cases of an engine-driven system, prime mover failure may be due to mechanical damage and continuous rotation by the generator may aggravate the situation. In a low head hydro turbine, continuous rotation may lead to cavitation of the blades.

Therefore, motoring protection is normally provided through reverse power relays. However, nuisance tripping may be caused by power swings and therefore a delay is introduced if a sensitive reverse power relay is used. The power required to motor a generator unit depends on its friction and may vary from 5% to 25% of the rated power. Therefore, the setting and the time delay should be set by considering the type of the prime mover and the reverse power that it draws.

4.4.1.5   Protection of mechanical systems

A generator may be provided with protection systems against over-speed, vibration, bearing failure, loss of coolant, loss of vacuum, etc. Some of these protection measures will only provide an alarm so that an operator may intervene to resolve the problem or initiate a manual shut-down. Some may provide backup protection. For example, if the vibration is caused by multiple earth faults on the rotor circuit, rotor earth-fault protection provides the main protection and vibration protection acts as a backup protection.

4.4.2   Protection of the faulted distribution network from fault currents supplied by the distributed generator

The common method of detecting a fault on the distribution network is by detecting the large fault current feeding into the fault using an over-current relay. However, in the case of a distributed generator, the fault current may be insufficient to be detected by an over-current relay. Therefore reduction in voltage, an increase or a decrease of frequency and/or a displacement of the neutral voltage are used to detect a fault on the distribution network fed by a distributed generator.

4.4.2.1   Over-current protection

Figure 4.14 shows a simple means of detecting phase faults. For large generators, this protection is employed as a backup protection for generator internal faults, as differential protection provides the main protection of the generator. For small generators (often less than 1 MVA [32]), this over-current protection provides the

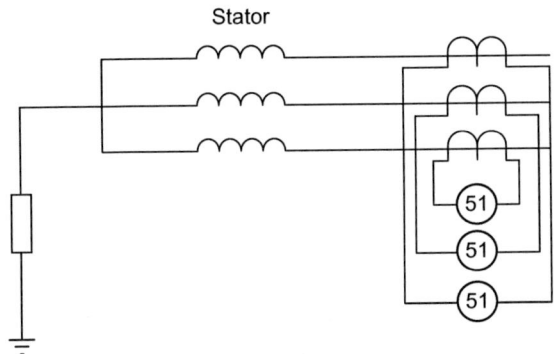

Figure 4.14   Over-current protection

main generator protection. It is also used to detect faults on the interface to the distribution network. To coordinate the operation of this relay with the downstream relays of the interface, sufficient delay should be introduced. Further, the pickup of the relay should be set to about 175% of the generator-rated current to accommodate transient currents due to a slow-clearing external fault, the starting of a large motor, or the re-acceleration current of a group of motors.

In order to obtain the required delay and pickup current, an inverse definite minimum time (IDMT) relay designated Type 51 is normally employed. This relay has two adjustments, namely [32]:

1. The pickup current setting of values between 50% and 200% in 25% steps. The 100% current setting normally corresponds to the rating of the relay.
2. The operating time can be adjusted by changing the time multiplier setting (TMS). TMS can be adjusted between 0.05 s (a smaller value of 0.025 is possible in most numerical relays) and 1.0 s.

Due to the high pickup current setting, this over-current scheme may not work for many faults on the distribution network. For example for a three-phase fault at F_1 on the network shown in Figure 4.8, the fault current provided by the distributed generator decays below the pickup level of the Type 51 relay in approximately 52 ms as shown in Figure 4.15. As the time delay of the Type 51 relay may be as high as 500 ms, the relay will not provide protection against three-phase faults.

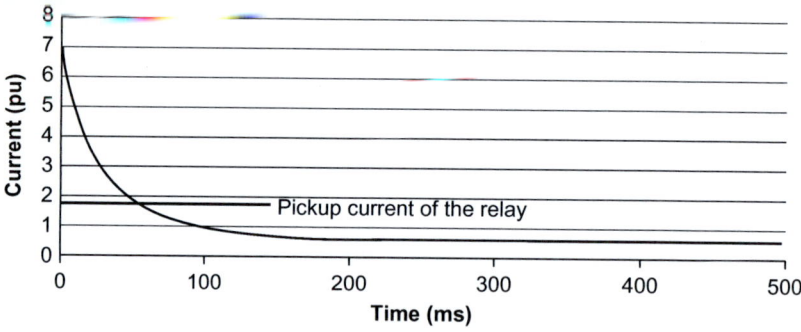

Figure 4.15   Decay of the fault current for a synchronous generator

Voltage-restrained or voltage-controlled time-over-current relays (Types 51VR, 51VC) may be used to improve the operation of over-current relays when fed from a distributed generator. The voltage-restrained and voltage-controlled time-over-current relays require both current and voltage signals to operate. The voltage feature allows selection of a pickup current below the rated current.

The voltage-restrained approach causes the pickup current to decrease with decreasing voltage as shown in Figure 4.16(a). When the voltage is at its rated value, the pickup setting of the relay is high ($I_{pickup1}$). As the voltage decreases (due

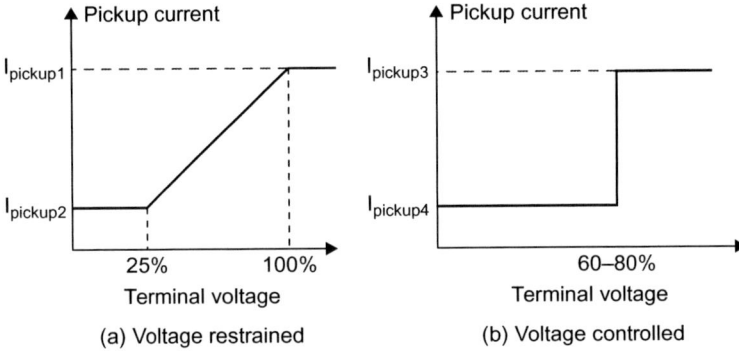

Figure 4.16   *Operating characteristic of voltage-restrained and voltage-controlled relays*

to a fault), the pickup value of the relay is decreased proportionally. For example, at a voltage of 25% or below, the pickup value drops to $I_{pickup2}$. The voltage-controlled approach has high pickup value ($I_{pickup3}$) when the voltage is above a preset voltage (60–80%) and the pickup current is reduced to a lower value ($I_{pickup4}$) for voltage below the preset voltage (Figure 4.16(b)).

For the network shown in Figure 4.8, a three-phase fault at F_1 reduces the voltage seen by relay C almost to zero. If a voltage-controlled over-current relay is employed, a pickup current ($I_{pickup4}$) of 50% of the rated may be set to operate the relay. During normal operation as the voltage is approximately 1 pu, the $I_{pickup3}$ can be set to a high value (e.g. 175%).

4.4.2.2   Under/over voltage and frequency

Abnormal conditions such as a fault in the power system, the disconnection of part or all of the network from the distributed generator (and its captive load), and rejection of a load or a generator may cause frequency or voltage excursion, which are considered to be outside the normal operating range defined in IEEE 1547 and G59/1. In such circumstances, under/over-voltage and under/over-frequency protection relays are used to disconnect the distributed generator from the network within the time specified in these standards.

Sustained over or under-voltage conditions are rare in a generator unless the AVR malfunctions or the unit is islanded. If the distributed generator is an induction generator without sufficient power factor correction, then under voltages can occur.

Under and over frequencies may occur in an islanded system due to generation–load mismatch. A power-system fault near the generator collapses the network voltage and so stops the generator exporting power. The generator then over speeds (as the input mechanical power remains unchanged), thus leading to over frequency and hence relay operation. This is often used to detect islanding.

4.4.2.3   Neutral voltage displacement

In an 11 kV network, which uses impedance earthing or which becomes unearthed due to islanding, a phase-to-earth fault increases the voltage of the system neutral point. For example, if the system is unearthed, then a phase-to-earth fault increases the neutral voltage (with respect to earth) to be equal to the phase-earth voltage. The earth-fault current flows only through the stray capacitance of the network and is not adequate to operate the over-current earth-fault protection. Therefore, neutral voltage displacement protection is used. The neutral voltage protection relay is normally connected across an open delta winding of a voltage transformer, where it receives zero voltage under normal operating condition. However, when an earth fault occurs the open delta voltage rises [32].

4.4.3   Anti-islanding or loss-of-mains protection

Generators connected to distribution networks and falling under IEEE 1547-2003 and G59/1 are not permitted to continue to operate if their connection to the main power system is lost. For these smaller generators, stability, neutral earthing and reliable fault currents are all provided by the main power system. Hence, protection is fitted to detect this loss of mains. The condition of no longer being connected to the main power system may be detected through: over/under voltage, over/under frequency, special loss-of-mains relays either detecting ROCOF (df/dt) or vector shift of the output of the islanded generator and detecting the opening of the connecting circuit breaker(s). This protection also inhibits the connection of distributed generation if the main power system (voltage and frequency) is not within its normal steady-state limits.

A relay operating by detecting the ROCOF may be used to provide protection against loss of mains. If the captive load in the islanded network is greater than the generator output, then the generator will slow down, thus seen as a reduction in frequency. The ROCOF is determined by the inertia of the distributed generator and captive load. Typically the setting of the ROCOF relay is set to about 0.1–0.2 Hz/s. However, a loss of a large generator on the system or the connection of a large load also causes a similar frequency excursion and may cause the ROCOF relay to operate.

A sudden connection or disconnection of a load and a fault in the network may cause a sudden shift in the terminal voltage vector of a distributed generator with respect to its normal operating voltage. As shown in Figure 3.1, if the generator is represented by a voltage behind an impedance, a change in current (say ΔI) will cause a sudden change of voltage given by $\Delta I \times (R + jX_s)$. Figure 4.17 shows the vector shift for a high impedance fault (occurring at F_1 of Figure 4.8).

Typically a vector-shift relay is set to operate for a change in angle of 6°. However in a weak network, the connection of large load or the disconnection of a large generator may cause a vector shift more than 6°, and a higher setting is used.

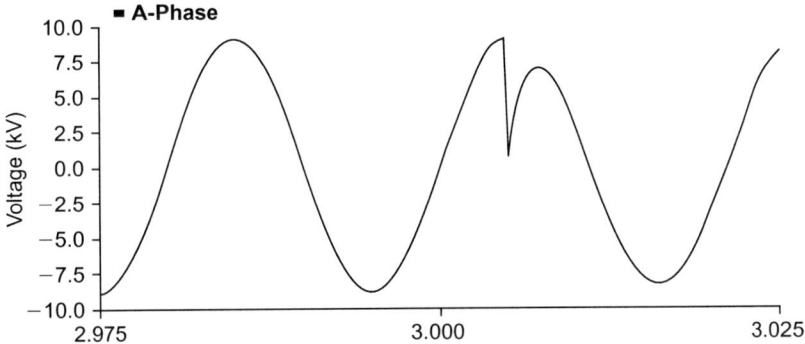

Figure 4.17   Vector shift created by a high impedance fault

4.5   Impact of distributed generation on existing distribution system protection

4.5.1   Phase over-current protection

Connecting a generator to the distribution network will effect the operation of existing over-current protection. For example, the additional fault current of a distributed generator may require changing the settings of the network over-current protection relays or the ratio of the CTs. The system shown in Figure 4.18 is used to explain this. The data for the lines, transformers, DG and load are given in the appendix.

In Figure 4.18, IDMT relays (OCR_1 to OCR_4) were originally set to provide discrimination without considering the distributed generator. Following conventional practice the most down steam relay is assigned the lowest TMS and then a grading margin[4] is used to calculate the settings of the other relays. More details of grading over-current relays in a distribution system can be found in References 32–34.

For a low impedance fault, the current flowing through relay OCR_1 remains unchanged after DG is connected. However, the fault current through OCR_3 and OCR_4 will increase due to the fault current contribution from the generator (see fault current paths A and B for a fault at F_1). From fault current calculations, it was found that the maximum fault current through OCR_3 was increased from 5569 A to 6527 A when DG is connected. This requires the CT associated with that relay to be changed from 300:5 to 400:5.[5] Once the CT is changed, the setting of the over-current relays should be adjusted to provide proper protection coordination.

---

[4] This is a time delay introduced to take into account the circuit-breaker tripping time and any errors in the relays.

[5] According to IEEE 242-2001, when fault current of more than 20 times the CT name-plate rating are anticipated, a different CT ratio should be selected.

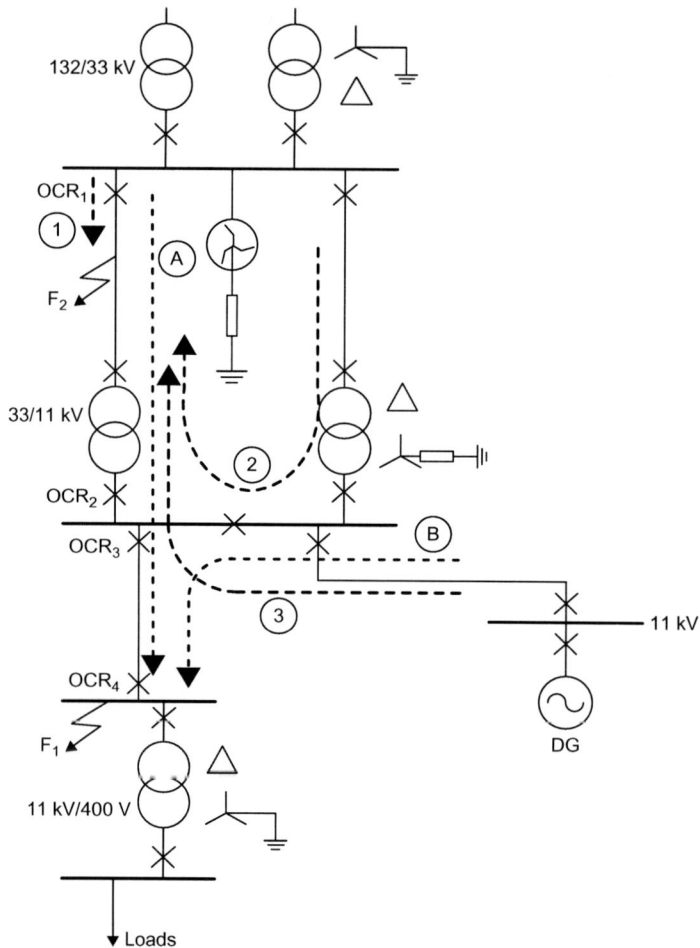

Figure 4.18   Effect of distributed generator on over-current protection grading

---

EXAMPLE 4.2

Part of a distribution system is shown in Figure E4.2. All calculations are carried out in per unit with $S_{base} = 100$ MVA and $V_{base} = 11$ kV.

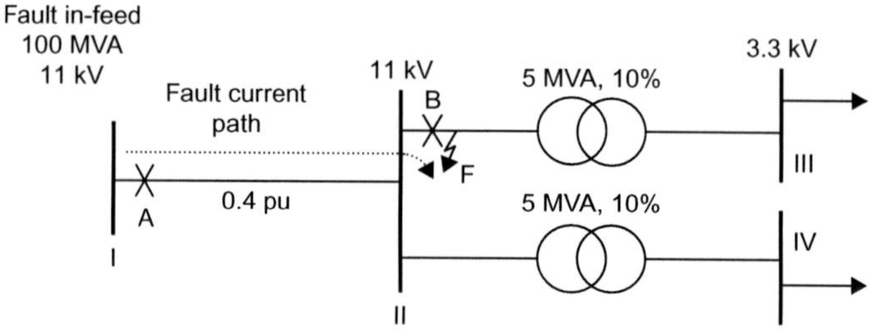

Figure E4.2

The 11 kV feeder (between busbar I and II) and the 11 kV/415 V transformers are protected using two IDMT relays (A and B). The relays are standard inverse type whose characteristic is defined by $t = 0.14 \times \text{TMS}/[\text{PSM}^{0.02} - 1]$ with the usual notation.[6] The CT ratios at A and B are both 200:5.

1. If the plug settings of relays A and B are set to 150% and the TMS of relay B is set to 0.05, calculate the TMS that should be set on relay A to achieve a grading margin of 0.3 s.
2. Now a generator of 4.5 MVA, $X' = 0.25$ pu (on the machine base) is connected to busbar IV.[7] What alterations are required on the over-current protection scheme to achieve correct grading?

1. $S_{\text{base}} = 100\,\text{MVA}, V_{\text{base}} = 11\,\text{kV}$ and so $I_{\text{base}} = \dfrac{100 \times 10^6}{\sqrt{3} \times 11 \times 10^3} = 5249\,\text{A}$

   The short-circuit current at bus II = $1/(1 + 0.4) = 0.708$ pu = 3749 A.

Relay	A	B
Fault current for a fault at F[8]	3749 A	3749 A
CT ratio	200:5	200:5
Plug setting	$1.5 \times 200 = 300$ A	$1.5 \times 200 = 300$ A
PSM (fault current/PS)	3749/300 = 12.5	3749/300 = 12.5
Operating time for TMS = 1	$t = 0.14 \times 1/[12.5^{0.02} - 1]$ 2.7 s	$t = 0.14 \times 1/[12.5^{0.02} - 1]$ 2.7 s

As the TMS on relay B is 0.05, the actual operating time of that relay for a fault at F = $0.05 \times 2.7 = 0.135$ s.

For proper grading, relay B should operate at least 0.3 s earlier than relay A for a fault at F.

Therefore, the actual operating time of relay A $\geq 0.3 + 0.135 = 0.435$ s.

As the operating time of the relay with TMS = 1 is 2.7 s, in order to achieve the above operating time TMS of relay A should be $\geq 0.435/2.7 = 0.16$.

Therefore, the value of TMS that should be set on relay A is 0.175.[9]

2. When the generator is connected to busbar IV, the fault current seen by relay B increases.

   The reactance of the transformer is $0.1 \times 100/5 = 2$ pu.

   The reactance of the generator is $0.25 \times 100/4.5 = 5.6$ pu.

---

[6] PSM is the ratio between the fault current and the plug setting.

[7] The transient reactance X' is used due to the delay in operation of the IDMT relays.

[8] Point F and busbar II are within the same substation and therefore the fault current is the same as that of Bus II.

[9] In a numerical relay, TMS can be set in steps of 0.025, and 0.175 is the nearest value to 0.16. If a lower value such as 0.15 is selected then the required grading margin cannot be achieved.

The per unit equivalent circuit with the DG is given in Figure E4.3.

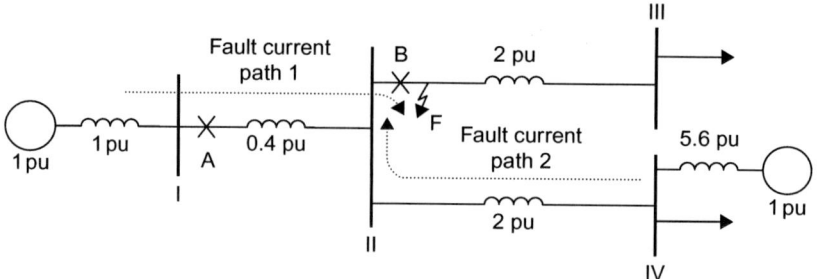

Figure E4.3

For a fault at F, the short-circuit current flows through two paths each having impedance of 1.4 pu (1 + 0.4) and 7.6 pu (2 + 5.6). When calculating the fault current, these impedances act in parallel giving an equivalent fault impedance of 1.18 pu. Therefore, the short-circuit current = 1/1.18 = 0.85 pu = 4440 A.

As fault current divided by 20 is greater than the CT rating at B, a higher rated CT should be selected (IEEE 242-2001).

Now select 300:5 CT at B. The new fault current at F and change in CT ratio demand recalculating the TMS of relay A to provide grading margin of 0.3 s.

Relay	A	B
Fault current for a fault at F	3749 A	4450 A
CT ratio	200:5	300:5
Plug setting	$1.5 \times 200 = 300$ A	$1.0 \times 300 = 300$ A[10]
PSM	$3749/300 = 12.5$	$4450/300 = 14.8$
Operating time for TMS = 1	$t = 0.14 \times 1/[12.5^{0.02} - 1]$ 2.7 s	$t = 0.14 \times 1/[14.8^{0.02} - 1]$ 2.5 s

The actual operating time of relay B for a fault at F = $0.05 \times 2.5 = 0.125$ s. The actual operating time of relay A $\geq 0.3 + 0.125 = 0.425$ s.

As the operating time of the relay with TMS = 1 is 2.7 s, in order to achieve the above operating time TMS of relay A should $\geq 0.425/2.7 = 0.157$.

The value of time multiplier setting of relay B need not to be changed.

---

[10] As the CT ratio is increased the same sensitivity as the previous case can be obtained by reducing the plug setting to 1.0.

4.5.2 Directional over-current protection

The 33 kV transformer feeder circuits in Figure 4.18 are provided with an over-current relay at OCR_1 and a directional over-current relay at OCR_2. OCR_2 was originally set to 50% (150 A). When a fault occurs at F_1 the fault current flows through two paths (1 and 2). The fault current through path 2 operates relay OCR_2 instantaneously. Then relay OCR_1 operates completely interrupting the fault. This will ensure uninterrupted operation of the healthy feeders.

With the distributed generator, the directional element may operate for reverse power flow in the transformer feeder during normal (unfaulted) operation. This will require an increase in the directional over-current setting. For example with a 7 MW generator and light load condition (10% of full load), the reverse current through OCR_2 is about 180 A. This demands an increase in the relay current setting, thus reducing the sensitivity of the relay to reverse faults currents (due to a fault at F_2).

4.5.3 Impedance relays

Distance relays are often used on 132 kV or 33 kV overhead lines, and a generator connected within the protected zone of these impedance relay may reduce the reach of the relay. The distributed generator acts to support the network voltage so appearing to increase the impedance to the fault seen by the relay. The relay then calculates that the fault is further away, outside the zone of protection and so will not operate. This can be explained using Figure 4.19.

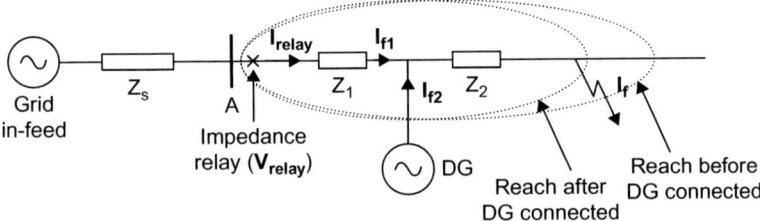

Figure 4.19   Effect of distributed generator on an impedance relay

Assume that before the distributed generator is connected, the setting of the impedance relay is $(Z_1 + Z_2)$. After DG is connected, the voltage seen by the relay and current through the relay for a fault within its zone of protection is given by:

$$\mathbf{V_{relay}} = \mathbf{I_{f1}} Z_1 + (\mathbf{I_{f1}} + \mathbf{I_{f2}}) Z_2 \qquad (4.10)$$

$$\mathbf{I_{relay}} = \mathbf{I_{f1}} \qquad (4.11)$$

Thus, impedance seen by the relay:

$$= \frac{\mathbf{V_{relay}}}{\mathbf{I_{relay}}} = Z_1 + \left(1 + \frac{\mathbf{I_{f2}}}{\mathbf{I_{f1}}}\right) Z_2 = (Z_1 + Z_2) + \left(\frac{\mathbf{I_{f2}}}{\mathbf{I_{f1}}}\right) Z_2 \qquad (4.12)$$

Since the impendence seen by the relay within the original protected zone of zone 1 is now increased above the setting of the relay, the impedance relay will not provide protection for the entire zone, thus under-reaching its protected zone.

A4.1   Appendix

Transformers:
  25 MVA, 132/33 kV transformer, reactance = 10.0%
  10 MVA, 33/11 kV transformer, reactance = 12.5%
  2.5 MVA, 11/0.4 kV transformer, reactance = 5.0%

Lines:
  33 kV line, length = 5 km and resistance = $(0.011 + j0.03)$ Ω/km
  11 kV line, length = 2 km and resistance = $(0.242 + j0.327)$ Ω/km
  Inter-tie of the DG, length = 4 km and resistance = $(0.242 + j0.327)$ Ω/km

DG:
  4.51 MVA, $X_s = 2.95$ pu, $X_d' = 0.25$, $X_d'' = 0.17$, $T_d' = 0.47$ and $T_d'' = 0.054$ [30]

Fault in-feed:
  900 MW

Load:
  1 MW at 0.9 pf lagging

References

1. Kimbark E.W. *Power System Stability – Synchronous Machines*. IEEE press; 1995.
2. Tleis N. *Power Systems Modelling and Fault Analysis*. Elsevier Press; 2008.
3. Kundur P. *Power System Stability and Control*. McGraw-Hill; 1994.
4. Hindmarsh J. *Electrical Machines and Their Applications*. Pergamon Press; 1970.
5. Krause P.C., Wasynczuk O., Shudhoff S.D. *Analysis of Electric Machinery and Drive System*. John Wiley Press; 2002.
6. IEC 60909-0:2001 (or BS EN 60909-0:2001). 'Short circuit current calculation in three-phase a.c. systems'. International Electromechanical Commission and/or British Standards Institution; 2001.
7. Electricity Association. 'Procedure to meet the requirements of IEC 60909 for the calculation of short circuit currents in three-phase AC power systems'. Engineering Recommendation G74; 1992.
8. Griffith Shan M. 'Modern AC generator control systems, some plain and painless facts'. *IEEE Transactions on Industry Applications*. 1976;**12**(6):481–491.
9. Grantham C., Tabatabaei-Yazdi H., Rahman M.F. 'Rotor parameter determination of three phase induction motors from a run up to speed test'.

International Conference on Power Electronics and Drive Systems; 26–29 May 1997, pp. 675–678.

10. Ao-Yang H., Zhe Z., Xiang-Gen Y. 'The Research on the Characteristic of Fault Current of Doubly-Fed Induction Generator'. *Asia-Pacific Power and Energy Engineering Conference*; 27–31 March 2009, pp. 1–4.

11. Morren J. 'Short-circuit current of wind turbines with doubly-fed induction generator'. *IEEE Transaction on Energy Conversion*. 2007;**22**(1):174–180.

12. Lopez J., Sanchis P. 'Dynamic behaviour of the doubly fed induction generator during three-phase voltage dips'. *IEEE Transaction on Energy Conversion*. 2007;**22**(3):709–717.

13. Anaya-Lara O., Wu X., Cartwright P., Ekanayake J.B., Jenkins N. 'Performance of doubly fed induction generator (DFIG) during network faults'. *Wind Engineering*. 2005;**29**(1):49–66.

14. CIGRE Working Group WG 37-23. 'Impact of increasing contributions of dispersed generation on the power system'. 23 September 1998.

15. Wall S.R. 'Performance of inverter interfaced distributed generation'. *IEEE/ PES Transmission and Distribution Conference and Exposition*; vol. 2, 28 Oct.–2 Nov. 2001, pp. 945–950.

16. Power A.J. 'An Overview of Transmission Fault Current Limiters'. Fault Current Limiters – A Look at Tomorrow, IEE Colloquium; June 1995, pp. 1/1–1/5.

17. Wu X., Mutale J., Jenkins N., Strbac G. 'An Investigation of Network Splitting for Fault Level Reduction'. Tyndall Centre for Climate Change Research Working Paper 25, January 2003.

18. Frank M. 'Superconducting Fault Current Limiters'. Fault Current Limiters – A Look at Tomorrow, IEE Colloquium; June 1995, pp. 6/1–6/7.

19. Yu J., Shi D., Duan X., Tang Y., Cheng S. 'Comparison of Superconducting Fault Current Limiter in Power System'. Power Engineering Society Summer Meeting; 1 July 2001, pp. 43–47.

20. Rasolonjanahary J.L., Sturgess J., Chong E. 'Design and Construction of a Magnetic Fault Current Limiter'. Ukmag Society Meeting; Stamford, UK, 12 October 2005.

21. Iwdiara M., Mukliopadhyay S.C., Yaniada S. 'Development of passive fault current limiter in parallel biasing mode'. *IEEE Transaction on Magnetics*. September 1999;**35**(5):3523–3525.

22. Putrus G.A., Jenkins N., Cooper C.B. 'A Static Fault Current Limiting and Interrupting Device'. Fault Current Limiters – A Look at Tomorrow, IEE Colloquium; June 1995, pp. 5/1–5/6.

23. Hojo M., Fujimura Y., Ohnishi T., Funabashi T. 'An Operating Mode of Voltage Source Inverter for Fault Current Limitation'. *Proceedings of the 41st International Universities Power Engineering Conference*; vol. 2, 6–8 Sept. 2006, pp. 598–602.

24. Chen M., Lakner M., Donzel L. Rhyner J., Paul W. 'Fault Current Limiter Based on High Temperature Superconductors'. ABB Corporate Research,

Switzerland. Article downloaded from http://www.manep.ch/pdf/research_teams/sciabb.pdf on the 01 November 2009.

25.   Rowley A.T. 'Superconducting fault current limiters'. *IEE Colloquium on High Tc Superconducting Materials as Magnets*; 7 December 1995, pp. 10/1–10/3.

26.   Paul W., Rhyner J., Platter F. 'Superconducting fault current limiters based on high Tc superconductors'. *Fault Current Limiters – A Look at Tomorrow, IEE Colloquium*; June 1995, pp. 4/1–4/4.

27.   IEEE 1547. 'IEEE standard for interconnecting distributed resources with electric power systems'. 2003.

28.   Engineering Recommendation G59. 'Recommendations for the connection of embedded generation plant to the public electrical suppliers distribution systems'. 1991.

29.   Engineering Recommendation G75/1. 'Recommendations for the connection of embedded generation plant to the public distribution systems above 20 V or outputs over 5 MW'. 2002.

30.   Engineering Technical Report E113. 'Notes of the guidance for protection of embedded generation plant up to 5 MW for operation in parallel with public electrical suppliers distribution systems'. Revision 1, 1995.

31.   IEEE 242-1968. IEEE recommended practice for protection and coordination of industrial and commercial power systems. *Color Book Series.* Green Book; 1986.

32.   *Network Protection and Automation Guide*, 1st edn. Areva T&D Ltd; 2002.

33.   Anderson P.M. *Power System Protection.* IEEE Press, McGraw-Hall; 1999.

34.   Bayliss C. *Transmission and Distribution Electrical Engineering.* Butterworth-Heinemann; 1996.

Chapter 5
Integration of distributed generation in electricity system planning

Prototype SeaGen marine current turbine, Strangford Lough, Northern Ireland (1.2MW)
The world's first grid-connected tidal turbine, installed and operational at Strangford Lough since 2008 – capturing energy from tidal flows [RWE npower renewables]

5.1 Introduction

For full integration of distributed generation, it is critically important to quantify its benefits and costs associated with electricity system development. As discussed earlier, distributed generation will displace not only the energy produced by central generation, but also the capacity of central generation and contribute to providing network security. In this context, this chapter describes the concepts and techniques needed for the impacts of distributed generation on the electricity system development to be determined, covering two key areas the:

- ability of distributed generation to displace the capacity of incumbent conventional generation and contribute to generation capacity adequacy
- ability of distributed generation to substitute for distribution network capacity and hence contribute to delivery of network security

5.2   Distributed generation and adequacy of supply

Sustainable energy systems with a significant penetration of distributed generation may require a diverse mix of generation technologies to maintain reliability of supply comparable to that of a conventional thermal system. Increasing the penetration of distributed generation will displace energy produced from existing plant, but it may not necessarily displace a corresponding amount of the generation capacity that is required to maintain an appropriate level of system reliability.

The impact of a generation technology on system reliability may be described by its capacity value or capacity credit. To calculate the capacity value of distributed generation technologies, the starting point is to establish the amount of conventional capacity required to maintain a given level of reliability of supply. Distributed generation is then introduced into the system, and an assessment is made to determine the amount of conventional generation capacity that can be retired to maintain the system security to its original level. This section discusses how the capacity value of distributed generation can be determined and how the corresponding cost or benefits can then be attributed to distributed generation.

5.2.1   Generation capacity adequacy in conventional thermal generation systems

In order to supply demand that varies daily and seasonally, and given that demand is at present largely uncontrollable while interruptions are very costly, the installed generation capacity must be able to meet the maximum (peak) demand. In addition, there needs to be sufficient capacity available to deal with the uncertainty in generator availability and unpredicted demand increases.

The planning of generation systems involves evaluating the generation capacity that is required to meet future demand with a certain level of reliability. The installed generation capacity should always be above the system maximum demand in order to deal with the uncertainties associated with both changes in demand and the availability of generation. Generation availability can be affected by forced outages of thermal generation plant and/or reduced water availability of hydro generation. The 'excess' of installed generation capacity above expected peak demand is called the capacity margin. Different capacity margins will deliver different reliability performance of the generation system.

The amount of generation capacity in a power system is considered to be adequate if it meets electricity demand with an 'economically efficient' level of reliability. Conceptually this 'optimal' level of capacity to be installed would be determined by balancing generation investment costs against benefits associated with the improvements in reliability of supply, i.e. the reduction in loss of supply to consumers. Instead of conducting such a cost–benefit analysis, generation system planners traditionally aim to maintain a certain level of capacity margin that will deliver a minimum reliability performance as measured by various reliability indices.

In a centrally planned system dominated by thermal generation, one of the frequently used indices is 'loss of load probability' (LOLP). This specifies the probability (or risk) that the system peak demand will be greater than the available generation (i.e. the probability that peak demand will not be met). The acceptable level of LOLP would always be below a certain threshold. In the last security standard employed in the United Kingdom, by the Central Electricity Generation Board ahead of privatisation in 1990, the risk of peak demand exceeding the available supply was required not to exceed 0.09, or 9% (interruptions in supply should not occur, on average, in more than nine winters in a century). Based on the probabilities of plant failure including uncertainty in the timely development of generation that corresponded to an equivalent generation availability of 85%, this standard would require a capacity margin of approximately 20%.

In determining the probability of various possible available outputs of a generator system, a standard two-state generator model was generally applied to simulate the behaviour of the generating unit. This assumes a generator unit is fully available, with the probability of 0.85, while also assuming that the generator unit will be completely unavailable, with the probability of 0.15. It was further assumed that there is no correlation between the availabilities of individual conventional units, i.e. the failure of one unit does not increase the risk of failure of others. Based on these assumptions it is possible to construct a generation system probability distribution that links the available level of the entire generation system output with the corresponding probability. We can also consider demand uncertainty through a probability distribution with a mean equal to the expected peak demand. This is shown in Figure 5.1.

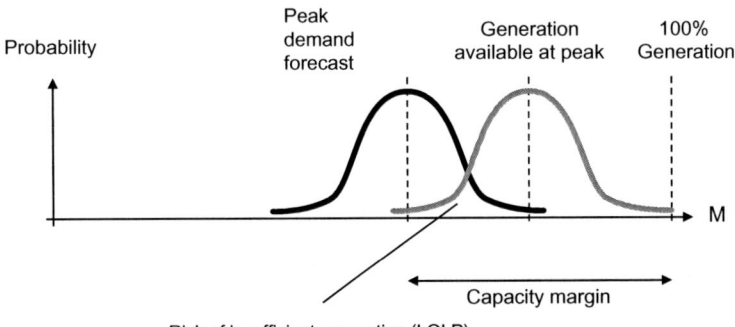

Figure 5.1 Probability distributions of demand and available generation

Figure 5.1 illustrates the relationship between the total capacity of installed generation and the system LOLP performance. It is clear that the higher the installed capacity of generation (higher capacity margins) the lower the LOLP (this moves the generation availability curve to the right and LOLP reduces). The generation capacity is considered adequate if the evaluated LOLP of a system (or future system in the planning domain) meets the threshold level.

This approach to planning and design of generation systems worked satisfactorily for several decades. Various reliability indices have been applied in different systems to evaluate adequate levels of generation capacity. More common indices include reserve margin, loss of load probability (LOLP), loss of load expectation (LOLE), expected energy not served (EENS) while less common indices include loss of energy probability, frequency and duration of failures, effective load carrying capability and firm equivalent capacity. It should be noted that the popularity of an index is not necessarily based on it giving a more accurate assessment of system reliability rather it was mostly due to the ease of its use and required input or data for its assessment.

Examples of applied reliability indices:

- The North American Electricity Reliability Council (NERC) reports the use of LOLP, LOLE and reserve margin to evaluate the adequacy of their regional generation portfolios. Many of the regional reliability councils (regions) in NERC's jurisdiction apply either LOLP (1 in ten years) or LOLE (0.1 day/year or 2.4 hour/year) [1].
- In Australia the reliability standard for generation and bulk supply are expressed in terms of the maximum permissible unserved energy or the maximum allowable level of electricity at risk of not being supplied to consumers. This is 0.002% of the annual energy consumption for the associated region or regions per financial year [2].
- France [3] and Republic of Ireland [4] apply LOLE criteria of 3 and 8 hours/year, respectively, to plan their generation systems.

These traditional planning approaches have been common practice for several decades in particular for a centrally planned thermal dominated system.

5.2.2 Impact of distributed generation

The methodology described in Section 5.2.1 can be extended to determine the adequate level of generation capacity in a system with wind, and other intermittent, generation. We primarily focus on the impact of wind generation, but the concepts are applicable to other forms of distributed generation. Unlike a two-state representation of conventional thermal generators, that are either in service and available to produce full output or not available at all (zero output), wind is modelled as a generator with multiple levels of outputs characterised by different probabilities. The frequency distribution of the aggregate wind generation can be obtained from the annual half-hourly profiles of wind output from historic wind output data.[1]

---

[1] This assumes that there is no correlation between peak demand and the level of wind output (which is generally conservative as wind farms in Europe tend to produce higher levels of output during winter peak periods).

The behaviour of conventional generation units and wind generation is then combined statistically, enabling the risk of loss of load (LOLP) to be calculated for systems that include both conventional and wind generation. This enables the capacity credit (value) of wind generation to be evaluated. The calculation starts with a conventional generation system that provides a reference level of LOLP. Wind generation is then added to the system, which will improve the reliability performance of the system (LOLP reduces below the reference value). We then determine the amount of conventional plant that can be removed from the system so that LOLP is equal to the reference level of the thermal only generation system. The ratio of the capacity of the conventional plant displaced by wind power to the total installed capacity of wind gives the capacity credit of wind.

The capacity credit changes with the levels of penetration of wind. Assuming a capacity factor of wind of 40%, Figure 5.2 illustrates the typical changes in the capacity credit of wind for different levels of its penetration. This is expressed as the installed capacity as a percentage of peak demand. The following observations can be made:

- The capacity credit reduces with wind penetration level.
- The capacity credit for low penetration levels approaches the load factor of wind.
- A diverse wind profile with wind farms occupying relatively large geographical area will exhibit higher capacity credit compared to a non-diverse profile with wind farms being concentrated in a small geographical area.
- The capacity credit of wind is greater than zero even though there will be finite probability of the wind output being zero during peak demand conditions.

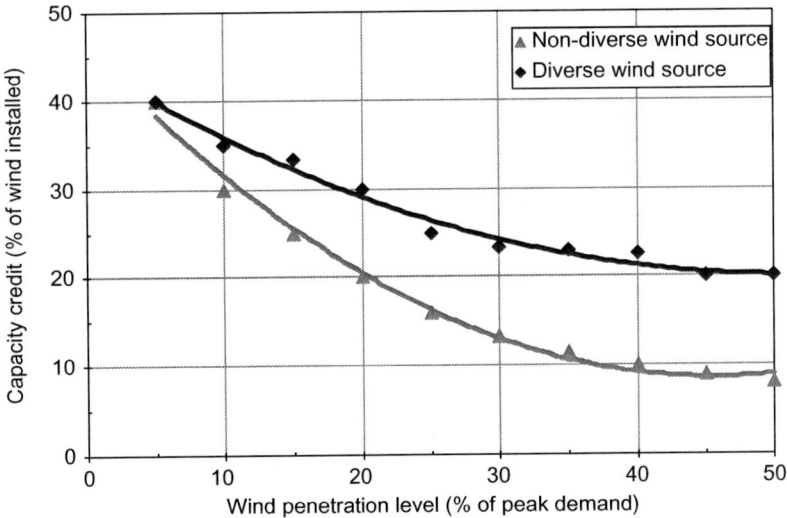

Figure 5.2   The capacity credit of wind generation versus the level of wind penetration in the UK electricity system

As wind generation has a relatively low capacity credit, it will displace proportionately more energy produced by conventional generation than the capacity of conventional generation. This will result in reduced utilisation of conventional plant. More generally, the relationship between energy and capacity displaced, and the resulting impact on average utilisation of the incumbent conventional generation, can be shown to be the primary driver for system costs associated with generation capacity:

$$\Delta C_{DG} = \left(1 - \frac{D_g^C}{D_g^E}\right) \cdot C_g^{I_o} \tag{5.1}$$

where ΔC_{DG} = additional system costs of distributed generation (£/MWh) that should be added to every megawatt hour produced by distributed generation (note that these costs could be positive and negative, depending on the actual technologies of distributed generation and the system), D_g^C = percentage displaced capacity of incumbent conventional technology (due to penetration of distributed generation technology), D_g^E = percentage displaced energy of incumbent (central) conventional technology (due to penetration of distributed generation technology) and $C_g^{I_o}$ = cost of capacity of the incumbent conventional generation expressed in pounds per megawatt hour. This cost is simply equal to the annuitised investment capacity cost of the incumbent conventional generation system (£/annum) divided by the total annual energy produced (MWh/annum). For a system based on CCGT, with a capacity of 84 GW (required to meet peak demand of 70 GW) generating 400 TWh/annum, the cost of generation capacity will be 14.07 £/MWh (with the annuitised capacity cost of CCGT of 67 £/kW/annum).

The values of these additional costs for wind generation of different characteristics, based on the UK wind generation profiles, are shown in Table 5.1.

Table 5.1   Additional system cost of wind power technology (£/MWh)

Load factor (%)	Capacity credit (%)			
	0	**10**	**20**	**30**
20	14.07	10.24	6.42	2.60
30	14.07	11.52	8.97	6.42
40	14.07	12.16	10.25	8.33

It can be observed that increase in capacity credit significantly reduces the additional system costs. We also see that an increase in load factor for a fixed capacity credit increases additional cost, which is caused by the reduction in the utilisation of incumbent central conventional generation technology. Taking an extreme position by assuming that wind has no capacity value the additional system cost of wind would amount to 14.07 £/MWh (this means that 14.07 £/MWh should be added to every megawatt hour produced by wind generation). It is important to mention that any base load plant that is added to the system (e.g. base load CCGT or nuclear) would also impose additional system costs. Such plant running at a load factor higher (say 85%) than that of average primary plant (say 55%) will displace more energy than capacity of the primary technology (similar to wind), causing a reduction in the (average) utilisation of the primary technology. This gives rise to an additional system cost of base load plant of about 5 £/MWh.

In contrast, micro-CHP output will typically be coincident with the winter peak demand, as this is coincident with the time of peak heating demands. Micro CHP will then effectively reduce the demand at the winter peak, thus reducing the capacity of conventional generation that is required to maintain system reliability. On the other hand the annual load factor of micro CHP will be lower than that of the average conventional generation. Hence, the conventional generation capacity displaced by micro CHP is greater than the displaced energy produced by conventional plant and the additional system cost in expression (5.1) is less than zero, indicating that micro CHP brings capacity benefits to the system.

In contrast to domestic CHP, PV generation installed in Northern Europe would have no capacity value as it cannot displace peak generation (e.g. it is dark at the time of GB peak demand, in the evenings of January and February). However, PV generation in Southern Europe, for example, where peak demand is in summer daytime, can contribute to reducing peak demand seen by conventional generation and hence bring benefits.

5.3   Impact of distributed generation on network design

At present distributed generation is not considered in the design of distribution networks. A simple example is set out in Figure 5.3 that illustrates how distribution network planners traditionally ensure security of supply of demand in a typical 33/ 11 kV substation design. As can be seen, the demand of 50 MW can be supplied through either of the distribution transformers such that if one circuit were to fail, the demand could be accommodated through the remaining transformer. In this example, the network planners would ignore the presence of the distributed generator and no security contribution is allocated to this resource.

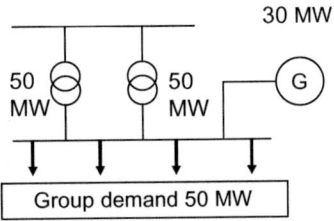

Figure 5.3   Example of secure network design without generation contribution

If the demand in this example were to grow to 55 MW, the supplying network would no longer be adequate, thus requiring the distribution network planners to seek some form of network reinforcements. Figure 5.4 illustrates two different approaches available to the planning engineers, which would secure sufficient additional capacity to make the network adequate. If the contribution of distributed generation is ignored, a third transformer would need to be installed to meet the security shortfall (Figure 5.4(A)). Alternatively, the distributed generator could potentially substitute for network reinforcement (Figure 5.4(B)).

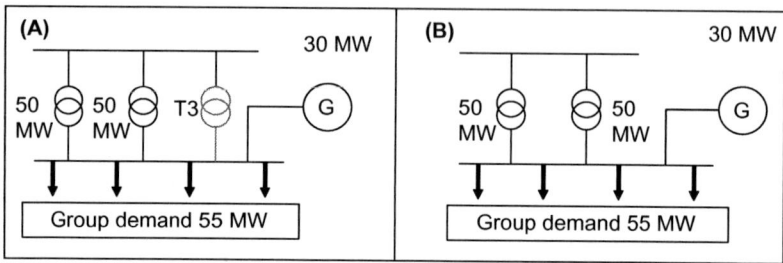

Figure 5.4   Network (A) and generation (B) solutions to a distribution network security of supply shortfall

The lack of an established framework by which distribution network planners can consider distributed generators for the provision of network security had been identified as a barrier to distributed generation. In the United Kingdom, the distribution network standards have changed recently to enable network planners to incorporate the contribution of distributed generation in network design. The concepts and techniques that can be applied to quantify the ability of distributed generation to displace network capacity and hence used in network planning are discussed in this section.

5.3.1   Philosophy of traditional distribution network planning

The UK network security standard defines the network design philosophy and the need to comply with it is a key network cost driver [5,6]. The level of security in distribution networks is defined in terms of the time taken to restore power supplies following a predefined set of outages. Consistent with this concept, security levels on distribution systems are graded according to the total amount of peak power that can be lost. A simplified illustration of this network design philosophy is presented in Figure 5.5. For instance small demand groups, less than 1 MW peak, are provided with the lowest level of security, and have no redundancy (n − 0 security). This means that any fault will cause an interruption and the supply will be restored only after the fault is repaired. It is expected that this could take up to 24 hours.

For demand groups between 1 MW and 100 MW, although a single fault may lead to an interruption, the bulk of the lost load should be restored within 3 hours. This requires the presence of network redundancy, as 3 hours is usually insufficient to implement repairs, but it does allow network reconfiguration. Such network designs are often described as providing n − 1 security. For demand groups larger

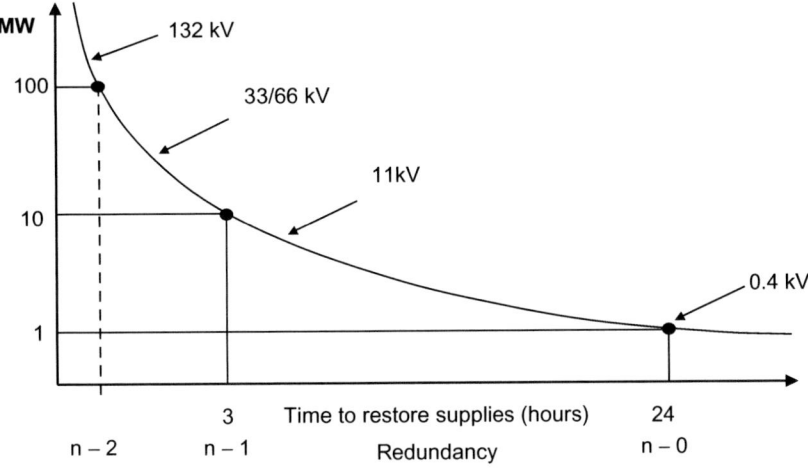

Figure 5.5 Restoration time philosophy relative to peak network demand

than 100 MW, the networks should be able to not only provide supply continuity to customers following a single circuit outage (with no loss of supply), but also provide significant redundancy to enable supply restoration following a fault on another circuit superimposed on the existing outage, i.e. n−2 security.

5.3.2 Methodology for the evaluation of contribution of distributed generation to network security

Distributed generation has challenged this conventional distribution network planning philosophy. In this section, we present concepts that have been applied in the United Kingdom to update network design standards to include effects of distributed generation. Given that the level of supply security provided by a network is measured through traditional reliability index such as expected energy not supplied (EENS), we examine the ability of distributed generation to substitute for network capacity through comparing EENS when supplied through a perfectly reliable network and a generator or a group of generators. This concept is shown in Figure 5.6.

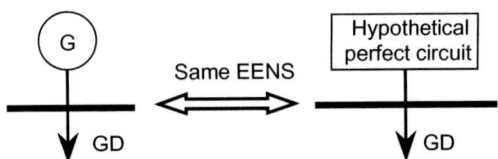

Figure 5.6 Comparing generation with a hypothetical perfect circuit.
Note: GD – group demand.

5.3.2.1   Treatment of non-intermittent distributed generation

The basic model for assessing the reliability of generation systems, the units of which are not constrained by intermittent energy sources and behave independently of each other, is best represented by a capacity outage probability table (COPT). The detailed theory relating to these is given in various reliability texts [7,8], but can be summarised as follows.

If all units in a given case are identical and behave independently, the capacity outage probability table can be evaluated using the binomial distribution in which the probability $P\{r\}$ of a specific state $\{r\}$ is given by:

$$P\{r\} = \frac{n!}{r!(n-r)!} p^r q^{n-r} \tag{5.2}$$

where n = number of units, r = number of available units, (n − r) = number of unavailable units, p = availability and q = unavailability of each unit.

If all units are not identical but still behave independently, the capacity outage probability table can be evaluated using the principle of state enumeration, i.e. if P_i and P_j are the probabilities of state $\{i\}$ and state $\{j\}$ respectively, then the probability of the combined state $\{ij\}$ is given by:

$$P_{ij} = P_i \cdot P_j \tag{5.3}$$

Consider the case of three identical units each having a capacity C and availability p. Using (5.1), the resulting capacity outage probability table is shown in Table 5.2.

Table 5.2   Capacity outage probability table for three identical units

Capacity available	Capacity unavailable	State probability
3C	0	p^3
2C	C	$3p^2(1-p)$
C	2C	$3p(1-p)^2$
0	3C	$(1-p)^3$
Total		1.0

In the case of three non-identical units having capacities C_1, C_2 and C_3 with availabilities of p_1, p_2 and p_3 respectively, the resulting capacity outage probability table is shown in Table 5.3.

Load is usually presented using the standard load duration curve (LDC), schematically presented in Figure 5.7. The specified period of time T, i.e. total horizontal axis, can be any time period of concern, e.g. one whole year, one season, one month, etc. The time units along the horizontal axis are usually hourly values. LDC represents the variation in load over a specified period of time in terms of the time duration the demand exceeds a particular load level (the demand exceeds load level L for t time units).

Table 5.3   Capacity outage probability table for three non-identical units

Capacity available	Capacity unavailable	State probability
$C_1 + C_2 + C_3$	0	$p_1 \cdot p_2 \cdot p_3$
$C_1 + C_2$	C_3	$p_1 \cdot p_2 \cdot (1 - p_3)$
$C_2 + C_3$	C_1	$(1 - p_1) \cdot p_2 \cdot p_3$
$C_3 + C_1$	C_2	$p_1 \cdot (1 - p_2) \cdot p_3$
C_1	$C_2 + C_3$	$p_1 \cdot (1 - p_2) \cdot (1 - p_3)$
C_2	$C_3 + C_1$	$(1 - p_1) \cdot p_2 \cdot (1 - p_3)$
C_3	$C_1 + C_2$	$(1 - p_1) \cdot (1 - p_2) \cdot p_3$
0	$C_1 + C_2 + C_3$	$(1 - p_1) \cdot (1 - p_2) \cdot (1 - p_3)$
Total		1.0

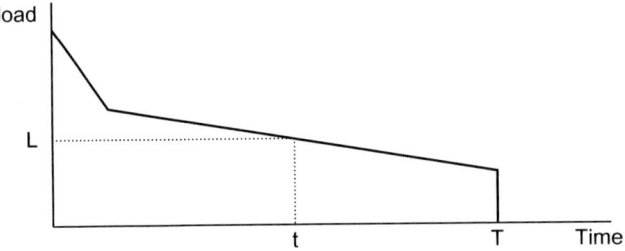

Figure 5.7   Simplified schematic shape of a load duration curve

Now, we can evaluate EENS for a group of generators supplying a load given by a load duration curve:

- each state of the capacity outage probability table is superimposed on the LDC individually as shown for one state *i* in Figure 5.8;

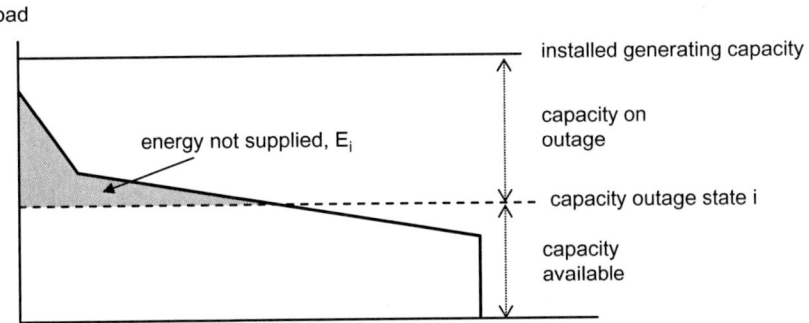

Figure 5.8   Evaluating EENS for a generation system

- the energy not supplied E_i whilst in this capacity state is determined as the area below the LDC and above the capacity available;
- this value of energy is weighted by the probability of being in this capacity state;
- these weighted values of energy are summated over all capacity states;
- from the concept of expectation, $EENS = \sum E_i.p_i$.

Finally, the capacity of the perfect circuit is calculated assuming that this capacity is constant, exists continuously and creates the same expected energy not supplied when this capacity level is imposed on the LDC. This capacity is defined as the effective output of the generation and the effective generation contribution is calculated as a ratio of effective output to maximum output.

This methodology can be used to determine the contributions that generating systems having different number of units and hence unit capacity (the total system capacity was kept constant), and different unit availabilities, make to system security. The results are shown in Figure 5.9.

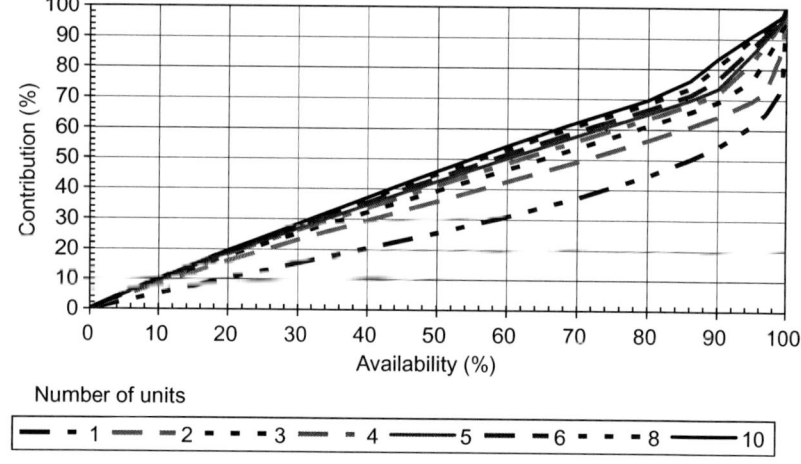

Figure 5.9   Effect of availability and number of units

As expected units with higher availabilities make larger contributions to security than units with lower availability. Furthermore, a single unit makes less contribution than a group of multiple units of equivalent capacity.

This analysis was the basis for the development of a new distribution network security standard that incorporates the contribution that distributed generation can make to network security. Table 5.4 is taken from the new standard to illustrate the features and the magnitude of the contribution to network security that is associated with various forms of distributed generation. The availability of individual units and the number of units are key driving factors of the contribution that distributed generation can make. The average values of actual availabilities for each type of non-intermittent plant are established and a security contribution evaluated. The table illustrates the 'F factors' for each characteristic technology; this represents a

Table 5.4   F factors in % for non-intermittent generation

Type of generation (*average availability, %*)	Number of generation units									
	1	**2**	**3**	**4**	**5**	**6**	**7**	**8**	**9**	**10+**
Landfill gas (*90*)	63	69	73	75	77	78	79	79	80	80
CCGT (*90*)	63	69	73	75	77	78	79	79	80	80
CHP										
Sewerage: spark ignition (*60*)	40	48	51	52	53	54	55	55	56	56
Sewerage: gas turbine (*80*)	53	61	65	67	69	70	71	71	72	73
Other CHP (*80*)	53	61	65	67	69	70	71	71	72	73
Waste to energy (*85*)	58	64	69	71	73	74	75	75	76	77

scaling factor to calculate the level output that can be 'guaranteed' against the total capacity of the generator.

5.3.2.2   Treatment of intermittent generation

The output level of intermittent generation changes frequently with time. The practical approach to deal with these variations is to characterise the variable as a time-varying parameter with its chronological behaviour fully represented. A schematic illustration of a generation pattern is shown in Figure 5.10. This chronological pattern captures all three types of availability: technical, energy and commercial availabilities, and hence is a suitable base model for representing the capacity states of the generation. Detailed knowledge of individual units, capacities, availabilities, etc. is therefore not required. In the case of multiple wind farms connected to the same demand group, the outputs of the individual farms should be aggregated and used as the chronological generation pattern. This enables diversity within and between sites, i.e. the footprint, to be taken into account.

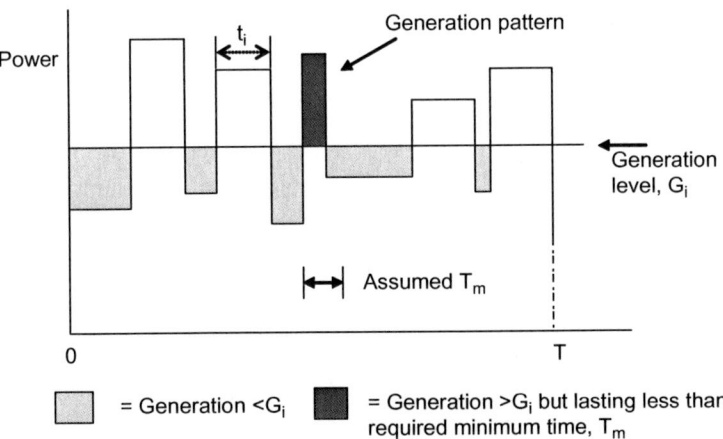

Figure 5.10   Determining model for intermittent generation

One requirement for security is that a particular generation output should be expected to remain at or above that level for the time period required. Therefore, the main difference between modelling non-intermittent and intermittent generation is those periods of intermittent generation that do not persist for a required minimum period of time must be discarded from the assessment. It is assumed that this minimum persistence time is T_m.

The steps in the assessment process are then:

- identify the time dependent generation pattern;
- consider a generation level G_i;
- identify the occasions when 'the generation is at least equal to G_i and continues to remain at least equal to G_i for the minimum time T_m';
- count the number of times, n_i, that this occurs and the duration, t_i, of each of these occasions;
- therefore, if T is the total time period of the generation pattern, the probability that the generation is at least equal to G_i is given by:

$$CP_i = \sum_i n_i \cdot \frac{t_i}{T} \tag{5.4}$$

- this can be repeated for all generation levels between the lowest and highest generation levels, from which a generation model can be determined. Each capacity state is given by G_i and the 'cumulative' probability by CP_i and all states are mutually exclusive;
- the individual state probabilities are determined from the cumulative probabilities. These states are imposed on the LDC in the same way as done for non-intermittent generation.

This procedure was implemented to determine the contribution to network security that can be expected from intermittent generation, such as wind and hydro. As shown in Table 5.5, the longer the time period over which a certain level of output is expected the lower the capacity contribution of the intermittent generator. For example, if the time over which the support is required is 30 minutes, 28% of the installed capacity of wind generation can be 'relied upon' while this value reduces to 11% for the time period of 24 hours.

Table 5.5   F factors in % for intermittent generation

Type of generation	Minimum persistence time, T_m (hours)							
	0.5	2	3	18	24	120	360	>360
Wind								
Single site	28	25	24	14	11	0	0	0
Multiple sites	28	25	24	15	12	0	–	–
Small hydro	37	36	36	34	34	25	13	0

5.3.3 Application of the methodology

The capacity of a network, i.e. the system capability, to meet a group demand should be assessed as:

- the appropriate cyclic rating of the remaining transmission or distribution circuits that normally supply the group demand, following outage of the most critical circuit(s);
- PLUS the transfer capacity that can be made available from alternative sources;
- PLUS for demand groups containing generation, the effective contribution of the generation to network capacity (as evaluated in Section 5.2.2).

The system used to illustrate the assessment of security is shown in Figure 5.11. In addition to the transformers, the system consists of two non-intermittent units, each with a declared net capacity of 20 MW and availability of 90%, and one wind farm of 10 MW.

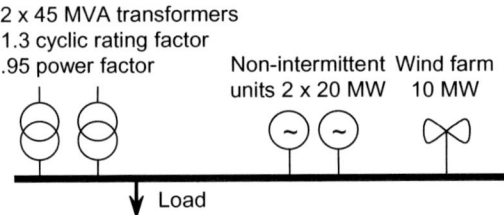

2 x 45 MVA transformers
1.3 cyclic rating factor
.95 power factor   Non-intermittent Wind farm
                         units 2 x 20 MW   10 MW

Load

Figure 5.11   System to illustrate evaluation of system capability

From the methodology described, the effective capability of each non-intermittent unit is 69%, and of the wind farm is 24% for $T_m = 3$ hours (see Tables 5.3 and 5.4). It is assumed that there is no available transfer capacity in this system, and that the most critical first circuit outage is the outage of one of the 45 MVA transformers. Hence, the overall network capacity is as follows:

- remaining circuit capacity following outage of the most critical circuit = 1 × 45 × 1.3 × 0.95 = 55.6 MW
- effective capability of generation = 0.69 × (2 × 20) + 0.24 × 10 = 30.0 MW

Hence the capacity to meet demand, including the contribution from distributed generation, is = 55.6 + 30.0 = 85.6 MW.

This principle can be extended to assess the system capability of any number of circuits, transfer capacities and generation combinations.

References

1. Resource and Transmission Adequacy Recommendations. North American Electric Reliability Council (NERC); 2004.

2. Comprehensive Reliability Review, AEMC Reliability Panel. Australian Energy Market Commission; 2007.
3. Generation Adequacy Report on the Electricity Supply-Demand Balance in France. RTE; 2009.
4. Generation Adequacy Report 2009–2015, EirGrid, Republic of Ireland. 2009. Available from http://www.eirgrid.com/media/GAR%202009-2015.pdf
5. ER P2/6 Consultation. Available from http://www.greenchartreuse.com/dcode/consultations.asp
6. ER P2/6 Security of Supply. Available from http://www.ena-eng.org/ENA-Docs/EADocs.asp?WCI=DocumentDetail&DocumentID=793
7. Billinton R., Allan R.N. *Reliability of Engineering Systems: Concepts and Techniques*. 2nd edn. New York: Plenum Publishing; 1992.
8. Billinton R., Allan R.N. *Reliability of Power Systems*. 2nd edn. New York: Plenum Publishing; 1996.

Chapter 6
Pricing of distribution networks with distributed generation

Wild Horse wind farm, Ellensburg, Washington USA (275 MW)
Blade delivery. Turbines are 2 MW variable speed using Doubly Fed Induction
Generators [RES]

6.1 Introduction

For distributed generation to compete successfully with central generation in a competitive environment, network pricing arrangements are critically important. Distributed generation, depending on the technology, operating pattern and exact connection point, may reduce the need for upstream network reinforcements and when located closer to the load, can reduce losses and potentially contribute to increasing local reliability of supply. The vehicle for realising this additional value is efficient network pricing.

In general, in a deregulated electrical power system, network prices should send signals to users reflecting the benefits they bring and costs they impose on network operation and/or development. Efficient pricing distinguishes between different locations and between different times of use, thus avoiding cross-subsidies and facilitating a level-playing field between distributed and central generation.

For distribution network with distributed generation, it is essential to take account of the contribution that the generation makes to network security, i.e. the ability of distributed generation to substitute for network capacity (as discussed in Chapter 5). Otherwise the value of distributed generation in this regard is not recognised and hence cannot be rewarded.

6.2   Primary objectives of network pricing in a competitive environment

Distribution network pricing has the following primary objectives.

Economic efficiency: There are essentially two types of costs: network operational costs and network development costs. In distribution systems short-term operational costs are at present dominated by the cost of distribution losses. In future the cost of network constraints may also need to be included. Network development costs involve investment in expansion or reinforcement of the network.

In a competitive environment, where coordinated generation and network planning is replaced by pricing, economic efficiency is achieved by sending price signals to users of the network so as to influence their decisions with regard to: (a) location in the network and (b) patterns of network use. This is the fundamental reason why economically efficient network use-of-system charges should be location and time-of-use specific. It is also worth noting that because the purpose of economic efficiency in pricing is to influence future behaviour, the investment costs that are relevant in the determination of efficient network use-of-system charges are the future network expansion costs[1] rather than past network development costs.

Signalling the need for future investments: Network pricing should send clear cost messages regarding the location of new generation facilities and loads as well as define the need for and location of new distribution network investments, i.e. encourage efficient network investment and discourage over-investment.

Deliver revenue requirements: Economically efficient prices based on network operating and/or development costs may not deliver the required revenue. These efficient prices then need to be modified to yield sufficient revenue to allow effective operation and timely development of distribution networks. This requirement may distort the objective of economic efficiency.

Provide stable and predictable prices: Price stability and predictability is important for users' investment decisions. The right balance must, however, be struck between price stability and flexibility allowing prices to respond to changing situations.

---

[1] The time horizon and assumptions on the locations of users, their future development and use patterns need to be defined for the future costs to be quantified.

Determination of prices must be transparent, auditable and consistent: The network cost allocation method should be transparent, auditable and consistent allowing users and other interested parties to understand easily the structure and derivation of network tariffs.

The pricing system must be practical to implement: Any proposed network pricing method should balance the economic efficiency of tariffs with their complexity as well as social objectives. From a practical standpoint the pricing method should be easy to understand and implement.

One of the major challenges in setting tariffs is establishing the trade-off between the various objectives of tariff setting. These include the ability to reflect accurately cost streams, efficiency in responding to changing demand and supply conditions, effectiveness in delivering appropriate revenue requirements, as well as stability and predictability of revenue and tariffs. It may be difficult to satisfy these all simultaneously.

6.3   A review of network investment cost drivers

The objective of distribution network pricing is to send signals to users on the costs they impose on network operation and/or development. Hence it follows that in order to calculate efficient network investment related prices, it is necessary to first establish future network investment costs. Future network investments and the associated costs are the outcome of network planning. Therefore, there is a close link between network pricing and network planning. In other words network planning is a key step in network pricing. Network prices then reflect the impact that individual users have on the costs of the planned network.

Network planning is driven mainly by planning standards (security and safety standards) and incentive mechanisms within the regulatory framework that may drive investment (quality of supply, losses and incentives to connect distributed generation). In general, the main investment drivers in distribution network design are as follows:

1. Network security: the need to satisfy network security requirements by investing in adequate network capacity
2. System fault levels: the requirement for adequately rated switchgear and network components
3. Network losses: the need to strike an optimal balance between operating costs and network investment
4. Service quality expenditure: the requirement to improve network performance indicators, e.g. in the United Kingdom, Customer Interruption and Customer Minute Lost indices

6.3.1   Network planning standards

A fundamental principle of distribution planning is that there should be sufficient capacity in the system such that, for predefined outage situations, customers continue to receive a supply or have it restored within an acceptable time. Furthermore, the network to which a distributed generator is connected should be able to absorb the full output of the distributed generation under all network loading conditions (here we assume uncongested distribution networks). Hence, in generation-dominated areas, the critical condition will be determined by the coincidence of maximum generation output and minimum load. If the network imposes no operational constraints under these conditions, all other conditions will be less onerous. In summary, two key loading conditions will need to be examined when determining the adequacy of the network capacity to fulfil its function:

1. Maximum load and minimum (secure) generation output
2. Minimum load and maximum generation output

With active network management of congested distribution networks, generators may decide to curtail their output in order not to drive investment.

6.3.2   Voltage-driven network expenditure

Distribution network operators (DNOs) have an obligation to supply their customers at a voltage within specified limits, and evaluation of voltage profiles under critical loading conditions is an integral part of network design. Sometime voltage drop (rise) is the key driver for network design and reinforcement.

In order to keep the voltage fluctuations within permissible limits, voltage control in distribution networks is carried out automatically by on-load tap changing transformers and sometime by reactive compensation installed at critical locations. For example, it is well known that the ratio of the MV/LV transformer is usually adjusted so that at times of maximum load the most remote customer receives acceptable voltage, just above the minimum value. On the other hand, during minimum load conditions the voltage received by all customers is just below the maximum allowed. The robust design of passive networks effectively minimises voltage variations across a wide range of operating conditions, e.g. from no load to full load.

Voltage considerations may determine the capacity (and hence resistance) of long distribution feeders, particularly in low- and medium-voltage networks. This is because the ratio of resistance over reactance of these circuits is usually significant, and the transport of active power in these networks has a significant impact on voltage. This is in contrast to high-voltage distribution and transmission circuits where reactive power flows determine the network voltage profile.

Therefore, when designing low- and medium-voltage circuits, in order to keep the voltage drop within limits, it may be necessary to select conductors of increased capacity, i.e. capacity that is greater than the necessary minimum dictated by thermal loading. In this case, voltage drop rather than thermal loading is the investment driver. It is however important to note that the maximum voltage drop will occur during maximum loading of the circuit. Hence, implicitly, maximum loading can be considered as the investment driver, given the voltage drop constraints. In the context of network pricing, we can treat loading as the primary cost driver of such circuits (bearing in mind that the actual capacity selected will need to be greater than the maximum flow to keep the voltage drop within allowable limits).

If a generator is connected to a demand-dominated area, its output offsets the network power flow and improves the voltage profile. The generator then reduces the demand for distribution network capacity and will postpone network reinforcement. This is very similar to a generator security contribution (discussed in Chapter 5) and the extent to which the generator offsets the downstream power flow (and hence improves the voltage profile).

6.3.3 Fault level driven network expenditure

All primary distribution plant transformers, cables, busbars and particularly circuit breakers and other switching equipment have a fault-level rating. The fault rating is not determined by the normal power the equipment can carry but by the maximum fault currents that the device is able to interrupt or pass. For circuit breakers a make and break capability is usually defined. Fault currents can be very significantly greater than the current under normal operating conditions.

In cost reflective pricing, it is possible to separate the allocation of cost of primary plant driven by normal loading, from cost driven by fault levels.

Many larger distributed generators use rotating machines and these, when connected directly to the network, will contribute to the network fault levels. Both induction and synchronous generators will increase the fault level of the distribution system. In urban areas where the existing fault levels approach the ratings of the switchgear, the increase in fault level can be a serious impediment to the development of distributed generation, as increasing the rating of distribution network switchgear and other plant can be very expensive.

A fault level analysis can be carried out to determine the contribution of each of the generators to fault levels at various busbars. This may then be used to allocate the cost of fault-level capacity to network users.[2] Distributed generators connected to the distribution network will contribute to fault levels but also connections that enter the distribution network at the grid supply points (as the majority of the fault current would normally come from the large conventional plant connected to the transmission network). This is illustrated in Figure 6.1.

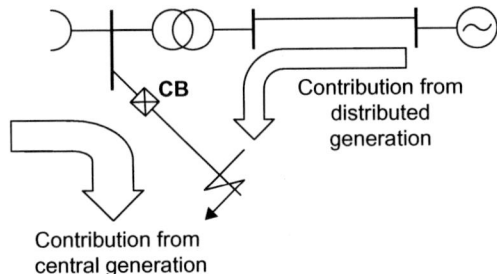

Figure 6.1   Contribution of central and distributed generators to fault level

---

[2] Fault-level analysis is carried out routinely during network design and can be implemented easily in the pricing exercise.

6.3.4   *Losses driven network design expenditure*

The impact of losses on the design of distribution cable and overhead networks can be profound. A minimum life-cycle cost methodology, which balances the capital investment against the cost of the system losses over the expected life of the circuit, can lead to very different capacities to those that result from a peak-load based network design.

In a minimum life-cycle cost methodology the optimal network capacity for transport of electricity is determined through an optimisation process where annuitised network capital costs and annual network operating costs (of which network variable losses are the most significant component) are traded off. This optimisation requires a calculation of the annual network cost of losses, and involves modelling of annual variations of load and generation as well as associated electricity prices including the mutual correlation between these quantities. The costs of network losses are then balanced against annuitised network capital costs to determine the optimal capacity required for the economic transport of electricity.

Results, given in the form of the optimal circuit utilisation under peak conditions (ratio of maximum flow through the circuit and optimal circuit capacity), for different voltage levels, and assuming 5% discount rate and 30-year asset life, are presented in Table 6.1 [1].

Table 6.1   Optimal utilisation of cables and overhead lines in a typical distribution network (expressed as a fraction of full-load rating)

	Type of conductor	
Voltage level (kV)	**Cable**	**Overhead line**
11	0.2–0.35	0.13–0.2
33	0.3–0.5	0.17–0.25
132	0.75–1	0.3–0.5

These results indicate that the optimal utilisation of distribution circuits, particularly at lower voltage levels, should be quite low. In addition, the optimal design of circuits taking the cost of losses into account satisfies the vast majority of security requirements at no additional cost, in cable networks up to 33 kV and overhead lines throughout distribution systems. This result is a combination of two effects. The first is the relatively large cost of losses due to the coincidence of high electricity price with high demand. The second is the relative fall in the price of cables and overhead lines due to maturity of the technology and increase in competition in the manufacture of this equipment.

As losses may drive investment in circuits, then charging for use of the network on the basis of peak demand as at present may not be appropriate as losses are present at all times. The question of the availability of distributed generation may be less important than is often thought if losses are considered to drive network design.

In this optimisation, the capacity determined for transport applies to cables and overhead lines only. Ratings of other items of plant such as transformers and circuit breakers are determined on the basis of other considerations.

6.4 Evaluating distribution use-of-system charges (DUoS charges)

In this section, we discuss the fundamental principles on which charges for the use of distribution networks with distributed generation can be evaluated. We first discuss the link between network planning and network pricing and then introduce two main approaches for evaluating network costs. It is also demonstrated that efficient network pricing will have time-of-use charges to reflect the fact that different operating conditions have different impacts on the design of individual circuits, in distribution networks with distributed generation. Finally we discuss the allocation of network costs showing that the users, given their pattern of operation and location, may tend either to increase the rating of an individual network circuit and hence they would pay for the use of that circuit, or tend to reduce the rating of the circuit in which case they would get paid for the use of the network.

6.4.1 Concepts of static and dynamic network pricing

The concept of a reference network is a construct derived from economic theory and has a long history [2–4], although the term 'reference network' (or 'economically adapted network') was not necessarily used explicitly by all of these authors. Nevertheless, the fundamentals of the idea of applying the 'global economic optimality' of the network for pricing purposes have been implicitly discussed. In particular, Farmer [3] pioneered the application of this concept for pricing transmission in a competitive environment. The application of the concept of reference networks to regulation of distribution networks is discussed by Strbac and Allan [5]. This approach has also been used to examine the relationship between short-term, locational marginal prices (associated with corresponding financial or physical transmission rights) and transmission investment [6–8].

The level of detail and hence the complexity involved in determining the 'global economic optimality' of a transmission and distribution networks may vary considerably. However, the reference network, in its simplest form, would be topologically identical to the existing network, with the generation and load layouts as in the real network, and would operate at the same voltage levels as the real one, but the individual transmission and distribution circuits would have optimal capacities, given load characteristics of all connected demand and generation. For each of the network investment drivers discussed in Section 6.3, two network planning and corresponding charging approaches can be considered.

1. Static, where the network capacity is evaluated from a green field network planning exercise that ignores the capacity of the existing network; the network is designed assuming a single, static, generation scenario and particular

demand profiles. An assumption would be made that the system would operate in perpetuity in this condition. The optimal network is then costed at the modern equivalent asset value, and the resulting charges can be interpreted to reflect the long-term cost that each particular user imposes on the network.

2.   Dynamic network development models, where the network evolves from the existing capacity of the present network over a specific time horizon; the network reinforcements would then be identified assuming an evolving generation background (including commissioning of new plant and closures of old plant) and evolving demand (load growth) across a number of years in future. During the simulation, the timing of future reinforcements of individual network circuits is then recorded. The annuitised net present value of all individual circuit reinforcements is then calculated and these future reinforcement costs (in an annuitised form) are then attributed to corresponding circuits to be allocated to users in different locations in relation to their impacts. This is based on the view that pricing should reflect future reinforcement costs that will be incurred: it is only future network costs that can be saved if demand/generation is not undertaken. The unused capacity or headroom of an individual circuit will be important when determining the time in the future when reinforcement is required. The larger the headroom, the further into the future reinforcement will be required. The resulting charges would then reflect the net present value of costs of all future network reinforcements that each particular user will impose on the network.

6.4.2   Time-of-use feature of DUoS charges in networks with distributed generation

In a network with distributed generation, it is necessary that in addition to maximum demand and minimum generation, the minimum load and maximum generation conditions need to be considered. This approach would lead to the application of time-of-use charges to reflect the fact that different conditions drive network design of individual circuits. Hence, the cost of distribution network circuits, through which the critical flows occur during peak demand periods (typically winter peak in Northern Europe), would be paid for during the peak demand periods only, while the cost of circuits through which the critical flows occur during the minimum demand periods (typically summer nights) would be paid for during minimum demand periods only. This would clearly result in not only location specific but also a time-of-use charging mechanism.

6.4.3   Allocation of network costs to network users

Once the critical flows are determined, optimal rating of all network circuits can be established (and timing of reinforcements in the case of dynamic pricing). The cost of the distribution network can then be allocated to each network user with respect to their contribution to driving the design of each individual circuit. For this conventional power flow, sensitivity analysis is applied. Users who, with their operating regime and location, tend to increase the rating of an individual circuit (i.e. their operation contributes to an increase in the critical flow), would pay for the use of that circuit, while users who tend to reduce the rating of the circuit (i.e. their operation reduces the critical flow) would get paid for the use of the network. Network charges then may be positive and negative, depending on whether the user pays or gets paid for the use of the network.

Given that the topology of typical distribution network is radial, this allocation can be quite simple. If the critical flow through a distribution network circuit is towards the grid supply point (i.e. towards the transmission network), that circuit is classified as generator dominated, meaning that all downstream generators should then pay for the use of this circuit during the period when the critical flow occurs. Downstream demand would hence get paid, as an increase in demand would reduce the critical flow through the particular circuit under consideration. On the other hand, when the critical flow on a particular circuit is downstream, this circuit is classified as demand dominated and hence demand pays for use of this circuit while downstream generators get paid.

6.5   Illustration of the principles of evaluating DUoS charges in networks with distributed generation

The principles for the evaluation of the use-of-system charges of distribution networks with distributed generation are shown on two examples. We first show how the concept of time-of-use, location-specific distribution network charges is applied on a simple, single voltage level network, consisting of two busbars. We then expand the consideration on a multi-voltage level radial network in which some of the circuit costs are driven by demand and some by generation, demonstrating how the efficient network pricing can be applied to a more complex distribution network.

6.5.1   Simple two bus example

The basic concept of allocating future network costs using marginal cost pricing is illustrated with the aid of two simple examples based on a two-bus system (see Figures 6.2 and 6.4). Figure 6.2 shows a small generator located at busbar 2, whereas in Figure 6.4 the generator size is increased. The two systems have identical demand levels and profiles for winter and summer periods. The annual demand profiles can be represented by a summer daily profile and a winter daily profile as shown in Figure 6.3. For simplicity we also assume that the duration of summer and winter periods is equal, each occupying 50% of a year.

The assumed annuitised investment cost for the circuit between busbar 1 and busbar 2 (for the both networks) is £7/kW/year.

We first need to establish the critical flow in this network that will be used for determining the rating of the circuit. Given the demand profile we need to consider four demand conditions and corresponding critical outputs from the distributed generator. These are presented in Table 6.2.

The loads and output of the distributed generator that would lead to the two critical loading conditions are:

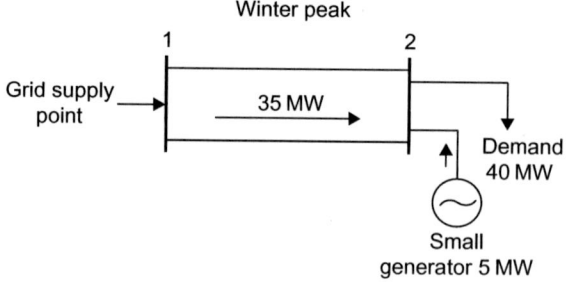

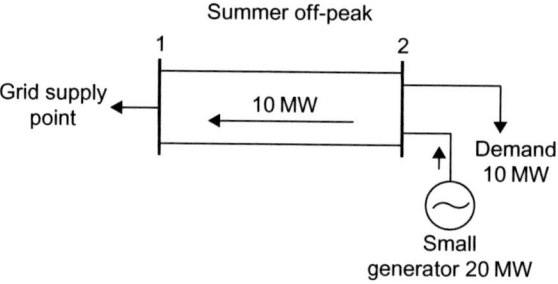

Figure 6.2   Two-bus system with a small generator

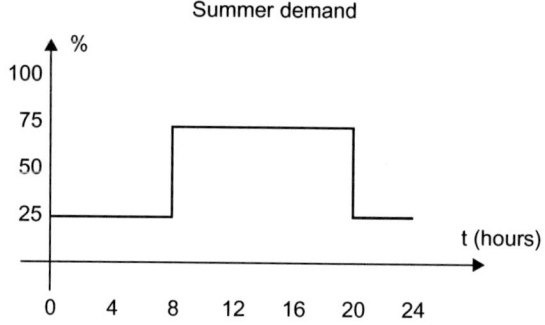

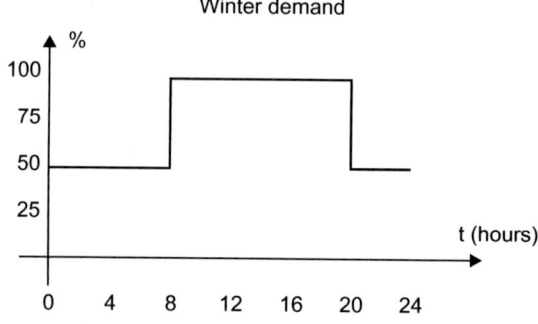

Figure 6.3   Demand profiles for Figures 6.2 and 6.4

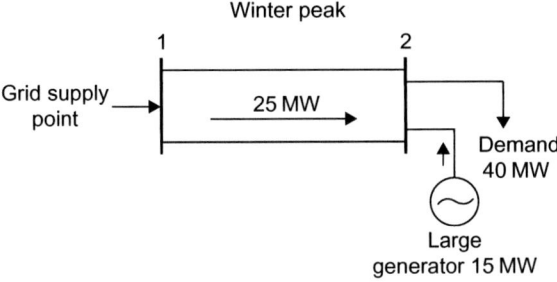

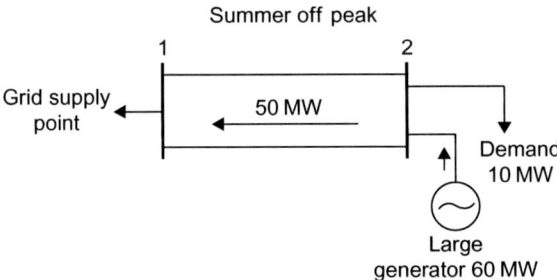

Figure 6.4   Two-bus system with a large generator

Table 6.2   Generation and demand profiles for two-bus system as shown
in Figure 6.2

		Winter peak	Winter off-peak	Summer peak	Summer off-peak	Critical flow	Classification of asset
Demand (MW)	A	40	20	30	10	–	–
Generation (MW)	B	5	20	5	20	–	–
Line flow (MW)	C = B − A	35	0	25	−10	35	DD

1. Maximum demand and minimum generation (for both summer and winter peaks). The generation output is equal to the generation contribution to network security, as discussed in Chapter 5.
2. Minimum demand and maximum generation (summer and winter off-peak). The generation output is equal to the capacity of the generator.

The peak and off-peak demand levels and the circuit flows are shown in Figure 6.2.

In Figure 6.2 the direction of the critical flow is determined by the demand, and hence the circuit is characterised as demand dominated (DD). In accordance with the concept of optimal costing, the use of the circuit between busbar 1 and busbar 2 is charged only during the period of maximum loading of the circuit. Clearly, if an additional unit of power flows through the circuit during its maximum loading (i.e. the winter peak period), this will require reinforcement of the circuit. When the circuit is loaded below its maximum capacity (all the remaining three periods), an additional unit of power will not lead to reinforcement, and hence the plant capacity charge is zero. Increasing demand at busbar 2 during the time of critical loading will require reinforcement of the circuit, hence demand should be charged for the use of this asset. On the other hand, increasing generation output during the critical loading condition (winter peak) will reduce the critical flow, thus benefiting the system. Generation should therefore be rewarded for conferring this benefit on the system.

A summary of the overall payments for the two-bus system in Figure 6.2 is given in Table 6.3.

Table 6.3   Nodal, time-of-use network price and user payments for system in Figure 6.2

		Winter peak	Winter off-peak	Summer peak	Summer off-peak	Total
Nodal DUoS at busbar 2 (£/kW)	A	7	0	0	0	
Demand level (MW)	B	40	20	30	10	
Generator output (MW)	C	5	20	5	20	
Demand payment (£'000)	D = A × B	280	0	0	0	280
Generator payment (£'000)	E = A × C	−35	0	0	0	−35
Net payment (£'000)	F = D + E	245	0	0	0	245

Note that negative payment for the generator in this case represents a revenue stream.

Consider now the same system, but with a larger distributed generator. We now repeat the same computations to find the critical flow, nodal prices and user payments with a larger distributed generator.

The resulting flows show that, unlike in the system shown in Figure 6.2 where the critical flow is determined by demand, the critical flow in this case is determined by generation. Hence, the circuit is classified as generation-dominated (GD) circuit (Table 6.4).

Table 6.4   Generation and demand profiles for the two-bus system shown in Figure 6.4

		Winter peak	Winter off-peak	Summer peak	Summer off-peak	Critical flow	Classification of asset
Demand (MW)	A	40	20	30	10	–	–
Generation (MW)	B	15	60	15	60	–	–
Line flow (MW)	C = B − A	25	−40	15	−50	50	GD

The computed nodal, time of use DUoS prices and associated user payments are shown in Table 6.5. Note that for consistency of the calculation of revenues and costs to different parties our nomenclature assumers that demand charges are positive and generation charges are negative.

Table 6.5   Nodal, time-of-use network price and annual user payments

		Winter peak	Winter off-peak	Summer peak	Summer off-peak	Total
Nodal DUoS at busbar 2 (£/kW)	A	0	0	0	−7	
Demand level (MW)	B	40	20	30	10	
Generator output (MW)	C	15	60	15	60	
Demand payment (£'000)	D = A ×B	0	0	0	−70	−70
Generator payment (£'000)	E = A × C	0	0	0	420	420
Net payment (£'000)	F = D + E	0	0	0	350	350

We now compare the two cases and Table 6.6 gives a summary of the final results from cases shown in Figures 6.2 and 6.4.

It is clear from Table 6.6 that the use-of-system charge is location and time-of-use specific:

- When the connected distributed generation is small, which leads to a demand-dominated circuit, the direction and the magnitude of the critical flow is determined by maximum demand. The nodal use-of-system charge is positive and demand pays for using the circuit. The demand charge applies in the winter peak period and is related to the peak demand. The generator gets paid according to its contribution to security (during the maximum demand period). In all other periods the DUoS charges are zero.

- When the connected distributed generation is large, the opposite situation applies. The system is generation dominated and the direction and magnitude of the critical flow is determined by maximum generation output and minimum demand. The nodal use-of-system charge is negative because the circuit is generator dominated and hence the generator pays for use of the asset, with respect to installed capacity (maximum generation), while demand gets paid with respect to minimum demand at the time of peak generation output, i.e. summer nights. In all other periods the DUoS charges are zero.

6.5.2   Multi-voltage level example

We now consider a radial distribution network composed of a 132 kV/33 kV substation with two transformers, two 33 kV out-going circuits (the rest of the 33 kV network is represented by a lumped load of 50 MW maximum demand), a 33 kV/11 kV substation with two 33 kV/11 kV transformers and two 11 kV feeders (the rest of the 11 kV network is represented by a lumped load of 10 MW maximum demand), each of which supplies four 11 kV/0.4 kV transformers with a maximum demand of 400 kW.

At the 33 kV busbar of the substation a 15 MW CHP plant is connected, and a wind farm of 1 MW is connected to one of the 11 kV circuits (Figure 6.5).

Table 6.6 Nodal network nodal, time use use-of-system charges and annual user payments

	Winter peak nodal price (£/kW) busbar 2	Summer off-peak nodal price (£/kW) busbar 2	Generator payment (price × generation in period) (£'000)	Demand payment (price × demand in period) (£'000)	Annual net payment (goes to pay for the asset) (£'000)
For the system shown in Figure 6.2	7	0	−35	280	245
For the system shown in Figure 6.4	0	−7	420	−70	350

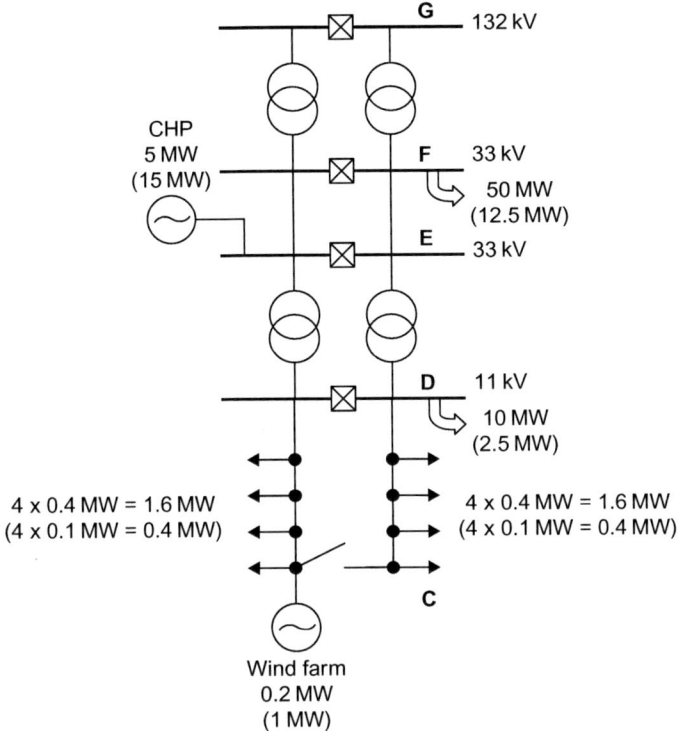

Figure 6.5   132/33 kV radial distribution system

A distribution network is typically designed to cope with the expected maximum loading condition, which is likely to occur at a time of maximum demand with minimum local generation. With distributed generation, another extreme condition needs to be considered, i.e. the condition where the generators produce maximum output and demand is minimum. Therefore, in this example, these two critical loading conditions are considered. The first number (without brackets) is the loading or generation during maximum demand–minimum generation (this would typically be associated with winter daytime), and the second number (in brackets) is the loading or generation during minimum demand–maximum generation (this would typically be associated with summer night time).

The design of the distribution network should take into account the contribution of distributed generation to network capacity. This was discussed in Chapter 5. For the sake of this illustrative example, an effective contribution of 5 MW (30% of installed capacity) is allocated to the CHP generator, and 200 kW (20% of installed capacity) to the wind turbine. In the context of DUoS pricing, it could be interpreted simply that the CHP and wind turbine are capable of substituting 5 MW and 200 kW respectively, of a distribution circuit capacity.

Minimum demand is assumed to be 25% of the maximum demand.

The flows in both loading conditions can be obtained by simple inspection. Critical flows are summarised in Figure 6.6. The arrows show the direction of the flows. The critical loading of items of plant can be seen to be the largest power flows of the two loading periods.[3]

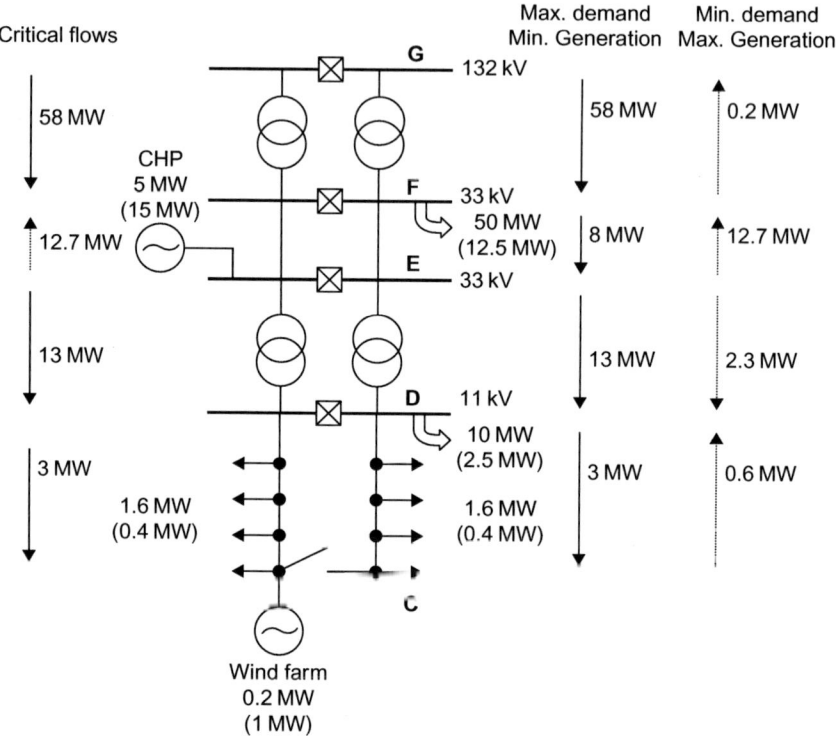

Figure 6.6   Computation of critical flows

It can be observed that critical loading for the 11 kV feeder, 33 kV/11 kV and 132 kV/33 kV transformers is driven by maximum demand, while critical loading of the 33 kV circuits is driven by maximum generation, and these occur at different periods (time of use). The assets whose capacity is driven by maximum demand–minimum generation condition (winter daytime) such as 11 kV feeders, 33 kV/11 kV

---

[3] Note that the critical flows determine the reference (optimal) ratings of the associated plant. The reference rating of 11 kV and 33 kV circuits are 2 × 3 MW and 2 × 12.7 MW respectively. Due to the topology of 11 kV circuits, one feeder must cope with all 11 kV loads when one of the 11 kV feeders loses supply from the 33 kV/11 kV substation and the normally open point is closed. The optimal rating of the 132 kV/33 kV substation is 2 × 58 MW. These reference ratings of the individual network components (transformers and lines at various voltage levels) can be compared with the plant ratings of the existing network.

and 132 kV/33 kV transformers, are then classified as demand dominated (DD), while the capacity of 33 kV feeders are driven by minimum demand–maximum generation condition (summer night time) and are hence classified as generator dominated (GD).

Given the direction of the critical flows and knowing the direction of demand- and generation-driven flows (Figure 6.6), it is easy to see how demand and generation at various voltage levels will pay or get paid for the use of individual network circuits. For example, an incremental increase in load of the demand connected to an 11 kV feeder, during the maximum demand periods, will increase the loading on the 11 kV feeder, 33 kV/11 kV and 132 kV/33 kV transformers. Therefore, this demand will be charged for the use of these circuits and the total charge will be based on maximum demand of 3.2 MW.

For the generation-dominated 33 kV circuit, the relevant critical period is determined by the coincidence of maximum generation and minimum demand. Hence, demand connected at 11 kV will be rewarded for the use of this 33 kV circuit, based on the load during minimum demand of 0.8 MW and the corresponding reward to demand will be obtained during the summer night periods.

Consider now charges for the wind farm. The wind farm will be rewarded for the use of the 11 kV network and 33 kV/11 kV and 132 kV/33 kV transformers and will be charged for the use of 33 kV circuits. The rewards for using these circuits will be based on the generator effective contribution to network capacity, i.e. 0.2 MW and apply during winter daytime. On the other hand, the charges for the use of 33 kV circuit will be based on the maximum generation output (1 MW) and applied during summer night periods.

In order to evaluate network charges for individual users, per unit annuitised capacity costs (£/kW/year) are allocated to each item of plant in the network. For illustrative purposes, the estimate annuitised typical costs of 132 kV circuits, 132 kV/33 kV transformers, 33 kV circuits and 33 kV/11 kV transformers for typical urban network in the United Kingdom are used.

The system is presented again in Figure 6.7 with all critical loadings highlighted.

The typical annuitised costs of individual circuits are shown next to the network model in Figure 6.7. DUoS exit charges for demand customers connected at various points in the network are also listed. The polarity of charges is adopted to be positive for downstream and negative for upstream power flows respectively.

Consider now the 132 kV/33 kV transformer. This is demand-dominated plant since the direction of the power flow is downstream. Hence, all downstream demand and generation customers pay and are paid 5.2 £/kW/year respectively for the use of this particular plant during maximum demand conditions, while charges are zero during the minimum demand period.

The next plant to be considered is the 33 kV circuit. This is a generation-dominated plant since the direction of the critical power flow is upstream. Hence, all downstream generation and demand customers pay and are paid 6.7 £/kW/year respectively for the use of this plant during maximum generation condition, while zero is charged during the maximum demand period.

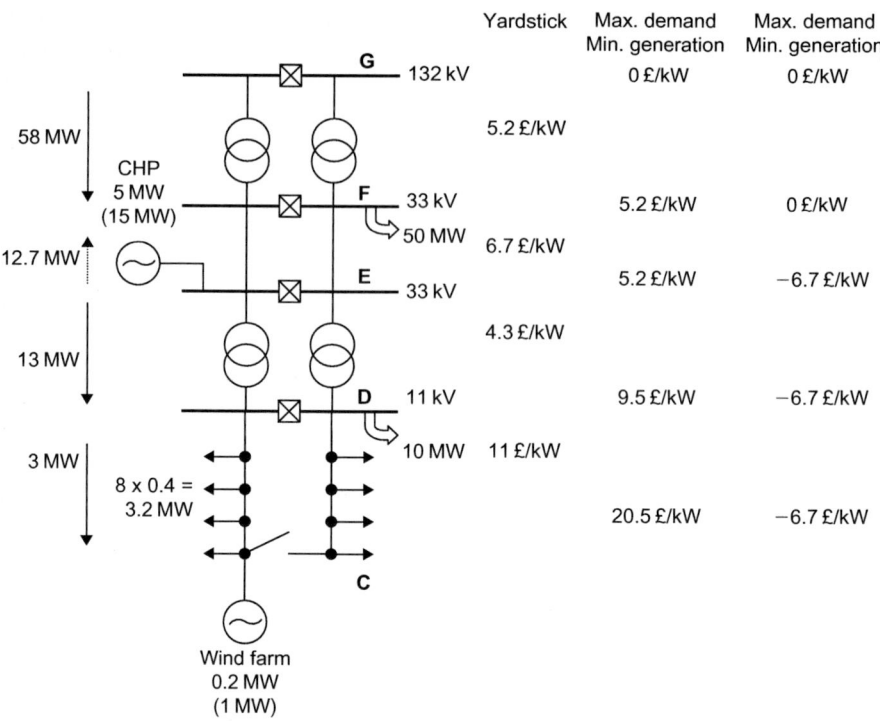

Figure 6.7   Evaluation of DUoS exit charges

As shown in Figure 6.7, the total DUoS exit charges for demand customers connected to the 33 kV busbar of the 33 kV/11 kV transformer is 5.2 £/kW/year applied during the maximum demand period (5.2 £/kW/year for the use of the 132 kV/33 kV transformer and 0 £/kW/year for the use of the 33 kV circuit), and DUoS entry charges of 6.7 £/kW/year during minimum demand period (0 £/kW/year for the use of the 132 kV/33 kV transformer and 6.7 £/kW/year for the use of the 33 kV circuit use).

The 33 kV/11 kV transformer is demand-dominated plant since the direction of the critical power flow is downstream. Hence, all downstream demand customers are charged and all downstream generation customers are paid 4.3 £/kW/year for the use of this particular plant during maximum demand conditions, while the charges are zero during the minimum demand period.

Therefore, the total entry charges for the generation connected to the 11 kV busbar of the 33 kV/11 kV transformer are −9.5 £/kW/year during maximum demand period (−5.2 £/kW/year for the use of the 132 kV/33 kV transformer, 0 £/kW/year for the use of the 33 kV circuit and −4.3 £/kW/year for the use of the 33 kV/11 kV transformer, and 6.7 £/kW/year during minimum demand period (0 £/kW/year for the use of the 132 kV/33 kV transformer) 6.7 £/kW/year for the use of the 33 kV circuit and 0 £/kW/year for the use of the 33 kV/11 kV transformer).

Finally, the 11 kV feeder is demand-dominated plant since the direction of the critical power flow is downstream. Hence, all downstream generation customers are paid 11 £/kW/year for the use of this particular plant during maximum demand conditions, while the charge is zero during the minimum demand period. Therefore, the total charge for generation customers connected to the 11 kV circuit is −20.5 £/kW/year during the on-peak period (−5.2 £/kW/year for the use of the 132 kV/33 kV transformer, 0 £/kW/year for the use of the 33 kV circuit, −4.3 £/kW/year for the use of the 33 kV/11 kV transformer and −11 £/kW/year for the use of 11 kV circuit) and DUoS entry charges of 6.7 £/kW/year during minimum demand period (0 £/kW/year for the use of the 132 kV/33 kV transformer, 6.7 £/kW/year for the use of the 33 kV circuit, 0 £/kW/year for the use of the 33 kV/11 kV and zero for transformer and for the use of 11 kV circuit).

The DUoS charges (assuming positive polarity for demand customers) and revenues collected from various users during peak demand and off-peak demand conditions are given in Tables 6.7 and 6.8. The connection point G corresponds to the balancing point. Note that 58 MW is imported under peak demand conditions while 0.2 MW is exported under minimum demand condition from the grid supply point (point G).

During the peak-load condition, the annual revenue is collected for all demand-dominated assets, while for the generation-dominated plant revenue is recovered

Table 6.7   On-peak demand DUoS prices and revenues from demand and generation customer

Connection point	Price (£/kW)	Demand (MW)	Generation (MW)	R demand (£)	R generation (£)	Total (£)
G	0	0	58	0	0	0
F	5.2	50	0	260000	0	260000
E	5.2	0	5	0	−26000	−26000
D	9.5	10	0	95000	0	95000
C	20.5	3.2	0.2	65600	−4100	61500
Total				420600	−30100	390500

Table 6.8   Off-peak demand (peak generation) DUoS prices and revenues from demand and generation customers

Connection point	Price (£/kW)	Demand (MW)	Generation (MW)	R demand (£)	R generation (£)	Total (£)
G	0	0	0.2	0	0	0
F	0	12.5	0	0	0	0
E	−6.7	0	15	0	100500	100500
D	−6.7	2.5	0	−16750	0	−16750
C	−6.7	0.8	1	−5360	6700	1340
Total				−22110	107200	85090

during peak generation periods. The costs of the individual plant items for the reference rating are given in Table 6.9.

In this particular case, the total annual revenue received for the demand-dominated plant is £390500/year, as shown in Table 6.7. (This is exactly equal to the total costs of the individual plant items as shown in Table 6.9, i.e. £390500 = £301600 + 55900 + 33000.) On the other hand, the total annual revenue received from DUoS charges during the off-peak demand period is £85090, as shown in Table 6.8. This is exactly equal to the total cost of generation-dominated circuit.

The on- and off-peak demand DUoS related expenditure of individual users is presented in Table 6.10. The total annual DUoS revenue equals the total annuitised cost of the reference network.

Table 6.9   Annuitised cost of individual plant items

Plant	Unit cost (£/kW)	Max. flow (MW)	Cost (£)
Transformer 132 kV/33 kV	5.2	58	301600
Circuit 33 kV	6.7	12.7	85090
Transformer 33 kV/11 kV	4.3	13	55900
Circuit 11 kV	11	3	33000
Total			475590

Table 6.10   Annual DUoS charges for individual network users

User	On-peak charge (£)	Off-peak charge (£)	Total charge (£)
Demand at F	260000	0	260000
Generator at E	−26000	100500	74500
Demand at D	95000	−16750	78250
Demand at C	65000	−5360	60240
Generator at C	−4100	6700	2600
Total			475590

References

1. Curcic S., Strbac G., Zhang X.-P. 'Effect of losses in design of distribution circuits'. *Generation, Transmission and Distribution, IEE Proceedings.* 2001;**148**(4):343–349.
2. Boiteux M. 'La tarification des demandes en pointe: applicationde la theorir de la vente au cout marginal'. *Revue General de Electricite.* 1949; **58**:321–340.
3. Farmer E.D., Cory B.J., Perera B.L.P.P. 'Optimal pricing of transmission and distribution services in electricity supply'. *Generation, Transmission and Distribution, IEE Proceedings.* 1995;**142**(1):1–8.

4. Nelson J.R. *Marginal Cost Pricing in Practice.* Prentice-Hall; 1967.
5. Strbac G., Allan R.N. 'Performance regulation of distribution systems using reference networks'. *Power Engineering Journal.* 2001;**15**(6):295–303.
6. Mutale J., Jayantilal A., Strbac G. 'Framework for allocation of loss and security driven network capital costs in distribution systems'. *IEEE Power-Tech International Conference on Electric Power Engineering*; Budapest 29 Aug–2 Sept 1999.
7. Mutale J., et al. *A Framework for Development of Tariffs for Distribution Systems with Embedded Generation CIRED'99.* 1999; NICE, France.
8. Mutale J., Strbac G. *Business Models in a World Characterised by Distributed Generation.* EC funded project number NNE5/2001/256 (April 2002 to March 2004).

Chapter 7
Distributed generation and future network architectures

Rhyl Flats Offshore Wind Farm (90 MW)
Situated five miles off the North Wales coast, in Liverpool Bay [RWE npower renewables]

7.1 Introduction

Distributed generation will continue to increase in importance as many countries move towards de-carbonising their energy systems. The location and size of many renewable energy sources and CHP plants require that they are connected to distribution networks and that effective use is made of existing circuits. This can only be done cost-effectively by integrating the operation of the distributed generation closely into that of the power system and changing the operation of distribution networks from passive to active [1–5].

Traditionally, power networks have been supplied from large, rotating synchronous generators that act as constant frequency and voltage sources. Hence, the entire operating philosophy of the power system has been developed around maintaining nearly constant voltage and, in the event of a short circuit, providing fault current to operate protective relays. A rather small number of these large rotating generators (some hundred units in Great Britain) are fitted with controls to maintain frequency, voltages and reactive power flows, ensure stability and provide damping of the power system.

Much distributed generation, either static or rotating, is connected to the network through power electronic converters via a DC link. In their simplest form, the power electronic converters simply inject real power at unity power factor into the network to maximise the output of the generator. They only provide close to load current into a network fault and the DC link decouples the generators from the network and so removes the effect of their spinning inertia.

The connection of generators via power electronic converters allows much greater flexibility and controllability than if a direct connection of a generator to the network is made. Hence these new forms of distributed generation, with their different operating characteristics, offer the possibility of new generator control philosophies as well as new system operating practices. The historical investment in the traditional power system is so great that its basic operating philosophy of constant voltages and frequency is most unlikely to change quickly, but already there is interest in using variable speed generators to provide injections of power at times of system frequency excursions using their stored kinetic energy as well as increased system damping [6].

The present limited penetration of distributed generation, operated to inject power with little consideration of the state of the power system, is already causing restrictions to the amount of renewable energy and CHP that may be connected. It is also leading to difficulties in the generation systems and transmission networks in some European countries, e.g. at times of high wind power and low load or during network disturbances. Therefore the present fit-and-forget philosophy, where distributed generators are viewed as negative loads and the distribution system is operated in the traditional manner, will be superseded by active integration of distributed generation through active network management. This will result in a blurring of the traditional distinction between transmission and distribution networks as distributed generation is controlled for the benefit of the power system by a distribution system operator. This gives particular challenges in Great Britain with its clear regulatory distinction between the suppliers of electrical energy and distribution network operators. The most effective way to coordinate and control a very large number of small generators and loads (perhaps up to 10^5 units) has yet to be determined although their aggregation into virtual power plants has been suggested.

The attractions of distributed generation to enable individuals to engage more closely with their energy supply should not be overlooked [7]. This is a rather fundamental question and may well become more pressing as climate change and energy security assume greater significance. It has similarities to the debate as to whether it is better for individual computers to hold their own software or to use it from a central source as required.

Hence, although there is widespread agreement that distributed generation must be integrated more effectively into the power system and the SmartGrid concepts should be adopted, there remains considerable ambiguity as to what this means in practice. In this final chapter, three strands of contemporary development are described:

- Active network management – which allows more distributed generation to be connected to distribution networks and operated effectively;
- Virtual power plants – which provides a means of aggregating a large number of small generators and facilitating access to markets;
- Microgrids – which allow the formation of small cells of microgeneration and controllable loads.

Techniques of active network management are being demonstrated on the public electricity system in a number of countries, while virtual power plants and microgrids are still the subject of research and demonstration projects.

7.2   Active network management

The transformation of the distribution network from passive to active operation has already started with demonstrations and early examples of the use of active network management. Although some of these techniques may seem rather obvious and straightforward, they allow considerable increase in the distributed generation that can be connected. The penalty is some increased complexity in distribution network control but this is manageable for individual schemes. A more serious difficulty comes with the dramatically increased complexity when a number of these individual active network management schemes are installed in the same section of network. The individual, ad-hoc active network management solutions are not coordinated with each other and their combined behaviour becomes difficult to predict. Then a system-wide solution becomes necessary. A number of trails and demonstrations are being developed [8–10] but so far there is no generally agreed approach to system-wide active network management.

7.2.1   *Generator output reduction and special protection schemes*

Figure 7.1 illustrates the significant benefit that can be obtained with a simple active network management scheme.[1] Two circuits, each of 10 MW capacity, are used to connect a distributed generator to the power system. The load at the busbar to which the generator is connected varies between 2 MW and 10 MW.

Under 'fit-and-forget' the generator must be able to export its full output at any time. Hence, assuming one circuit can be out of service at any time, the maximum rating of generation that could be connected is 12 MW (10 MW export to the network plus 2 MW minimum load).

If the generator is a wind farm, it will only operate at its rated output for less than, say, 30% of the time, depending on the wind speeds. These times of full output

---

[1] This illustration is based on MW flows with no consideration of voltage or power factor.

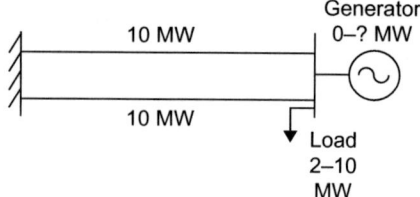

Figure 7.1   Illustration of simple active network management

are unlikely to coincide often with times of minimum load and so a wind farm capacity of greater than 12 MW, up to 20 MW, can be installed and the generation managed actively. The generation is then operated with either the power flows in the connecting circuits or in the load monitored. If excessive flow in the connecting circuits (more than the 10 MW firm capacity) is detected, the wind farm output is reduced. The choice of wind farm size is based on a cost-benefit calculation to determine the most cost-effective capacity of wind farm considering the wind resource and load, both of which can be estimated over a long period with confidence.

A development of this generator output control concept is to monitor the state of the connecting circuits and to use more of their thermal capacity. Here it is assumed that normally both circuits are in service, giving a capacity of 20 MW. This would then allow a wind farm of between 22 MW and 30 MW to be connected, depending on the degree of curtailment of generator output that is considered acceptable. If one of the circuits trips or is out of service, the wind farm output is reduced to a maximum of 10 MW plus the load. The special protection scheme trips the circuit for a fault and immediately reduces the output of the wind farm.

It may be seen through these simple examples that the increase in capacity of distributed generation which may be considered for connection is significant. Of course, issues such as voltages, stability and protection need detailed study and may well become limiting factors. However, experience has shown that significantly increased capacities of generation may be connected, particularly on MV circuits of 33 kV and above, if active management of this type is used.

A key administrative barrier is the basis on which the offer of connection is made to the developer of the generation scheme. Once the fit-and-forget philosophy is abandoned, the generator no longer has access to the network for all of its output, all of the time. The connection of a larger generator, with some restriction on its operation, may well be in the developer's commercial interest, but there is a degree of risk introduced as to how much energy it will be able to export. This may lead to difficulties in financing the project. It is also likely that electrical losses in the connecting circuits will be increased as the circuits become more heavily loaded.

7.2.2   Dynamic line ratings

Another active network management approach is to monitor the environmental conditions of the overhead lines and increase their rating when possible. In Northern Europe, high wind speeds, and hence full output of wind farms, tend to be during the winter months when the thermal capacity of overhead line circuits is increased by the low ambient temperatures and increased wind speeds over the conductors. These ambient conditions are monitored and used to calculate the capacity of the overhead line (particularly the sag of the conductors) to allow increased current to flow.

7.2.3   Active network voltage control

On medium voltage overhead networks, particularly 11 kV in the United Kingdom, the limiting factor for the connection of generation is often steady-state voltage rise. This is caused by the real power generated acting on the resistance of the circuit, while its low reactance means that absorbing reactive power is not an effective way of controlling the voltage rise. Hence, ways to control the tap changers of the 33 kV/ 11 kV transformer in order to increase the amount of distributed generation that may be connected have been investigated.

Figure 7.2 shows one approach [11]. Measurements of the network are taken (voltage magnitudes and power flows) and supplied to the distribution management system controller (DMSC) to determine the optimum operating points of both the tap changers and the generator outputs. The value of the energy from the various generators is also communicated to the DMSC, which in this example is located at a 33 kV/11 kV substation. Although control of the generators is considered by the controller, either by varying their power factor or reducing their real power output, the most cost-effective control action is likely to be changes in the set-point of the transformer tap changers. Loads are only shed in extreme situations given the much higher costs of shedding load compared to operating a tap changer.

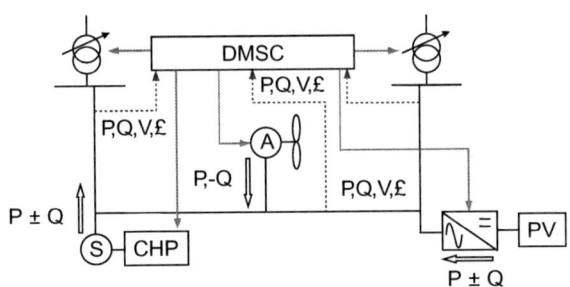

Figure 7.2   Active distribution management

Figure 7.3 shows how real-time measurements from the network may be combined with historical load data in an under-determined distribution state estimator to give a representation of the voltage magnitudes within the network. These voltages and power flows are then used by a controller, either using a simple voting system or an optimal power flow algorithm, to determine the best control action to be taken [11,12].

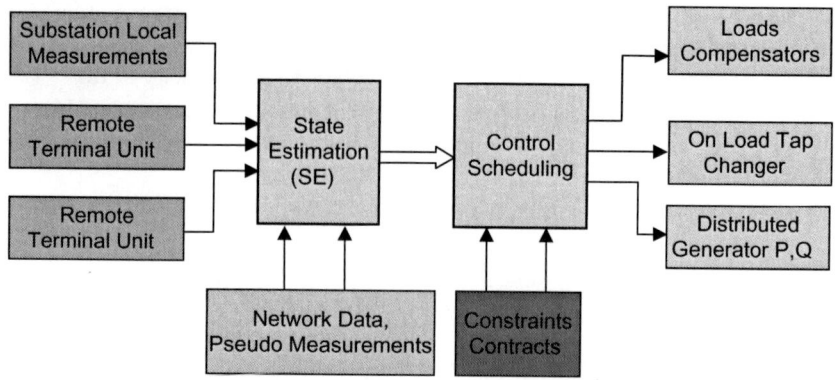

Figure 7.3   Distribution management system controller

7.2.4   Integrated wide-area active network management

More comprehensive wide-area active network management schemes have been investigated in a number of research and development projects [13]. An example is shown in Figure 7.4 where a controller is located at each 33 kV/11 kV substation not only to control its 11 kV network, but also to communicate with adjacent substations and to the higher voltage levels.

7.2.5   Smart Metering

The introduction of Smart Metering to all customer loads with data available in real time has the potential to increase dramatically the visibility of voltages and power flows within the distribution system. At present, there are very limited real-time

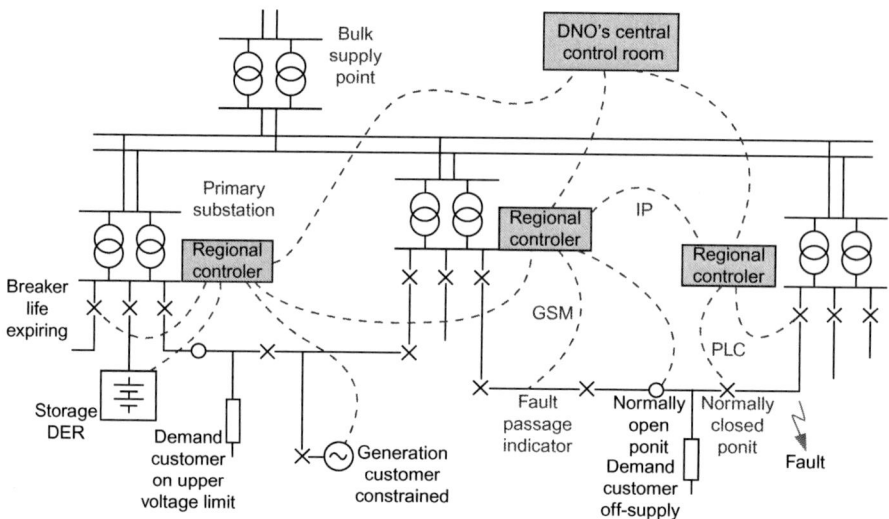

Figure 7.4   Wide-area active network management [13]

measurements on the distribution network and so its state can only be estimated using historical load data and the limited number of measurements that are present. Once the data from the Smart Meters becomes available it will, in principle, be possible to use the type of state estimator found on the transmission network (over-determined with more measurements than states) to give a much more robust and accurate picture of the distribution system.

7.3   Virtual power plants

To date, distributed generation has generally been used to displace energy from conventional generating plants but not to displace their capacity. Small distributed generators are not visible to system operators and are controlled to maximise energy from renewable sources or in response to the heat needs of the host site and not to provide capacity for the power system. Continuation of this way of operating the power system will lead to very large generation plant margins, under-utilisation of assets and low operating efficiencies. The concept of virtual power plants has been developed to increase the visibility and control of distributed generation, and to allow very large numbers of these small units to be aggregated so that they can take part in the various markets for energy and ancillary services [14].

In a virtual power plant (VPP), distributed generators together with responsive loads are aggregated into controllable units. These aggregated groups of generators are visible to the power system operator, can be controlled to support system operation and can trade effectively in energy markets. In short, they behave on the power system in a manner similar to that of large transmission-connected generation.

Through aggregation into a VPP

- individual distributed generators can become visible, gain access to energy markets and so maximise revenue opportunities.
- system operation benefits from effective use of the capacity of distributed generators and increased efficiency of operation.

7.3.1   The virtual power plant

When operating alone many distributed generators do not have sufficient capacity, flexibility or controllability to allow them to take part effectively in system management and energy market activities. A virtual power plant (VPP) is a representation of a portfolio of distributed generators and a vehicle through which small generators can take part in power system operation. A VPP not only aggregates the capacity of many diverse distributed generators, it also creates a single operating profile from a composite of the parameters characterising each small generator. The VPP is characterised by the set of parameters usually associated with a traditional transmission-connected generator: scheduled output, ramp rates, voltage regulation capability, reserve, etc. Furthermore, as the VPP also includes controllable loads, parameters such as demand–price elasticity and load recovery patterns are also used to characterise the VPP. A virtual power plant performs in a manner similar to a transmission connected large generating unit (Figure 7.5).

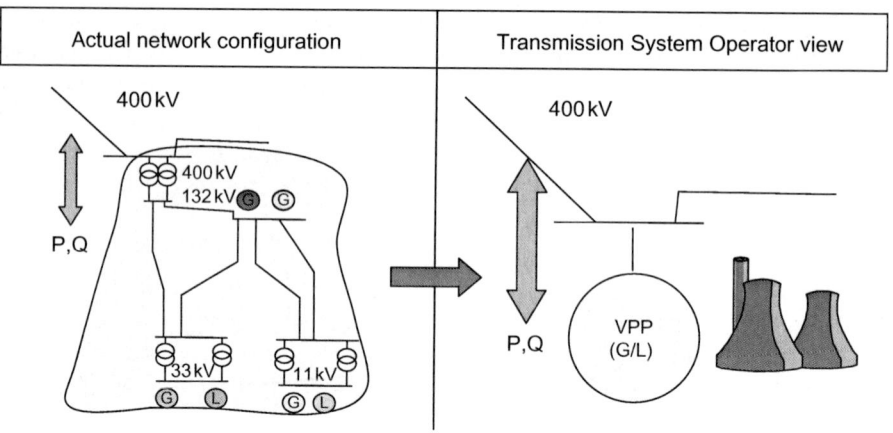

Figure 7.5   Aggregation of generators (G) and loads (L) into a virtual power plant [14]

Table 7.1 outlines some examples of generator and controllable load parameters that can be aggregated and used to characterise the VPP.

Table 7.1   Examples of generation and controllable load parameters that are aggregated to characterise a virtual power plant

Generator parameters	**Controllable load parameters**
• Schedule or profile of generation	• Schedule or profile of load
• Generation limits	• Elasticity of load to
• Minimum stable generation output	energy prices
• Firm capacity and maximum capacity	• Minimum and maximum
• Reserve capacity	load that can be
• Active and reactive power loading capability	rescheduled
• Ramp rates	• Load recovery pattern
• Frequency response capability	
• Voltage regulating capability	
• Fault-level contribution	
• Fault ride through characteristics	
• Fuel characteristics	
• Efficiency	
• Operating cost characteristics	

As a VPP is composed of a number of distributed generators, of various technologies operating patterns and availability, the characteristics of the VPP may vary significantly over time. Furthermore, as the generation of a VPP is connected to various points in the distribution network, the network characteristics (network topology, impedances and constraints) will also impact the overall characterisation of the VPP.

The VPP can be used to facilitate trading in the wholesale energy markets, but can also provide services to support transmission system management through, for example, various types of reserve, frequency and voltage regulation. In the development of the VPP concept, these activities of market participation and system management and support are described respectively as 'commercial' and 'technical' activities, which define the two roles of commercial VPP (CVPP) and technical (TVPP).

7.3.2   The commercial virtual power plant

The commercial VPP (CVPP) is a representation of a portfolio of distributed generators and controllable loads that can be used to participate in energy markets in the same manner as transmission connected generating plant. For distributed generators in the portfolio, this approach reduces imbalance risk associated with lone operation in the market and provides the benefits of diversity of resource and increased capacity achieved through aggregation. Distributed generation can benefit from economies of scale in market participation and market intelligence to maximise revenue.

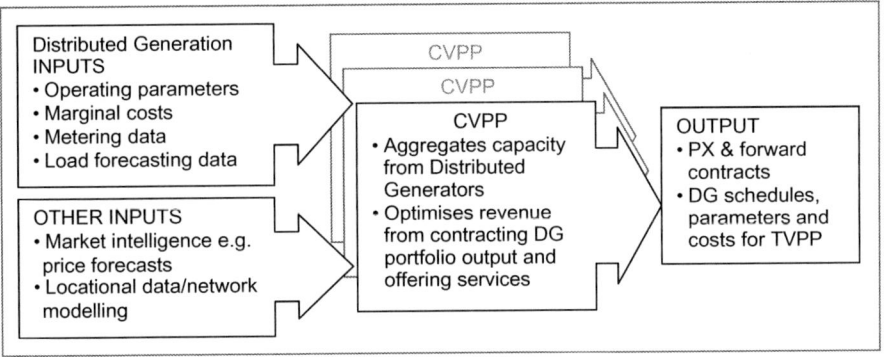

Figure 7.6   Inputs to and output from a CVPP

Figure 7.6 shows the CVPP inputs and outputs, each distributed generator that is included in the CVPP submits information on its operating parameters and marginal cost characteristics. These inputs are aggregated to create the single CVPP profile representing the combined capacity of all distributed generators in the portfolio. With the addition of market intelligence, the CVPP will optimise the revenue of the portfolio making contracts in the power exchange (PX) and forward markets, and submitting information on the distributed generation schedules and operating costs to system operators.

In systems allowing unrestricted access to the energy markets, i.e. with no network constraints, CVPPs can aggregate distributed generation from any geographic location. However, in markets where energy resource location is critical, the CVPP portfolio will be restricted to include only generators and loads from the same location (e.g. distribution network area or transmission network node). In these instances, a VPP can still represent distributed generation from various places, but aggregation of resources must occur by location, resulting in a set of generation and load portfolios defined by geographic location. This scenario may be expected in, for example, (transmission system) locational marginal pricing based markets and in markets where a zonal approach is taken to participation.

A CVPP can include any number of distributed generators and individual generators, and loads are free to choose a CVPP to represent them. The commercial VPP role can be undertaken by a number of market actors including incumbent energy suppliers, third party independents or new market entrants.

7.3.3   The technical virtual power plant

The technical VPP provides visibility of distributed generation to the system operator(s); it allows distributed generation to contribute to system management and facilitates the use of controllable loads to provide system balancing at lowest cost.

The technical VPP aggregates controllable loads, generators and networks within a single electric geographical area and models its response. A hierarchy of TVPP aggregation may be created to characterise systematically the operation of distributed generators connected to the low-, medium- and high-voltage regions of a local network, but at the distribution–transmission network interfaces the TVPP presents a single profile representing the whole local network.

Figure 7.7 summarises the inputs and output of a TVPP; information on distributed generation in the local network is passed to it by the various CVPPs. In the local distribution network, distributed generation operating positions, parameters and bids and offers collected from the CVPPs can be used to improve visibility to the distribution system operator and to assist with real-time network management and the provision of ancillary services.

To facilitate the contributions of distributed generation to operation at the transmission level, the TVPP operator (usually the distribution system operator) aggregates the operating positions, parameters and cost data from each distributed generator together with detailed network information (topology, constraint

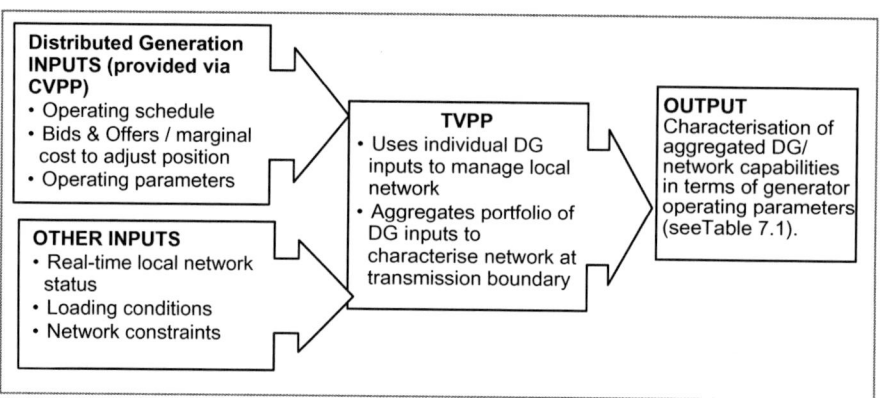

Figure 7.7   Inputs to and outputs from a TVPP

information, etc.) and forms the TVPP characteristics. The TVPP is defined at its point of connection to the transmission system, using the same parameters as transmission-connected plant (e.g. as outlined in Table 7.1). This TVPP profile and marginal cost calculation (reflecting the capabilities of the entire local network) can be evaluated by the transmission system operator along with other bids and offers from transmission-connected plants, to provide real-time system balancing.

Operating a TVPP requires local network knowledge and network control capabilities. Hence, the distribution system operator (DSO) is likely to be best placed to carry this role. With this TVPP capability, the DSO role can evolve to include active management of the distribution network, analogous to that of a transmission system operator. It is likely that the DSO would continue to be a local monopoly and any additional active management responsibilities would be regulated activities.

It is important to recognise that the market and regulatory framework surrounding distribution network operation and development is still evolving and that increasing penetration of distributed generation will lead to further unbundling and decentralisation of system management functions. This will affect the way in which the concepts of CVPP and TVPP are implemented.

7.4 MicroGrids

Microgrids have been investigated in a number of national and international research and demonstration projects with considerable work undertaken on both their design and operation. The research has yielded much valuable information as to how this element of a SmartGrid might be realised but there is, as yet, no general agreement on either the best microgrid architecture or the control techniques that should be used. To date, microgrids have not been implemented widely in public electricity supply systems, and a robust commercial justification or business case for their use has yet to be developed [15–18].

A microgrid can be defined as an electrical network of small modular distributed generators (microgenerators). Their prime movers are typically photovoltaic systems, fuel cells, micro turbines or small wind generators. The microgrid also includes energy storage devices and controllable loads. Most of these microgenerators are connected to the microgrid circuit through power electronic converters.

A microgrid can operate in grid-connected or islanded mode, and hence increase the reliability of energy supplies by disconnecting from the main distribution network in the case of network faults or reduced power quality. It can also reduce transmission and distribution losses by supplying loads from local generation and form a benign element of the distribution system.

Operating a microgrid in islanded mode has very considerable technical challenges, as normally the main power system provides to grid-connected microgenerators

- a well-defined frequency and voltage reference;
- a reliable and predictable source of short-circuit current;
- a sink or source of real and reactive power if the local load does not instantaneously match the microgeneration.

Microgrids lack a number of attributes around which large central power systems have been developed. On a high-voltage transmission system, the reactance (X) of the circuits is much greater than the resistance (R). This allows effective decoupling of control of the flows of real power (P) and reactive power (Q). If the resistance of a circuit is ignored (a reasonable assumption for the high-voltage network), it can be assumed that reactive power flows in a circuit are controlled by the magnitudes of voltages at each end of the circuit and real power flows are determined by the relative angles of the voltages. P and Q flows can then be considered independently. This simplifying assumption cannot be made on a microgrid where the resistance of the circuits may exceed their inductive reactance.

Frequency in a high-voltage power system is determined by the balance of generation and loads either taking kinetic energy from the spinning generators and loads or by supplying energy that accelerates the spinning machines. The speed with which the system frequency changes, when a large load is connected or a generator trips, is determined by the total inertia of all the spinning machines on the system. In a microgrid, many of the microgenerators either produce DC or are connected through power converters that decouple the generator from the AC voltage. If the loads are static (or connected though converters), then no spinning masses are directly connected to the microgrid circuits and so in islanded operation the frequency must be synthesised through the converter control systems.

In all the microgrid experiments conducted to date, it has been found necessary to use electrical energy storage to ensure stable operation when the microgrid is disconnected from the distribution network and to accommodate load changes when the microgrid is operating in islanded mode. Both flywheels and battery energy storage systems have been used [18–20]

Recently microgrid research has been extended to consider networks of other energy carriers (heat and gas systems fed from biogas). The objective is to supply energy to urban areas in an integrated manner and so reduce carbon emissions (Figure 7.8).

7.4.1   Microgrid research and demonstration projects

There have been three main strands of microgrid research, in the European Union, the United States and Japan [15–17].

In the European Union, two major research projects on microgrids have been undertaken. The structure proposed by the EU MicroGrid projects is shown in Figure 7.9 and the test system in Figure 7.10 [18]. In these laboratory microgrids, PV generation, fuel cells and micro turbines were interfaced through inverters, and small wind generators connected directly. A central flywheel energy storage unit was installed to provide frequency stability and short-circuit current during islanded operation. Islanding was done by disconnecting at medium voltage in order to retain the neutral earth connection of the MV/LV transformer.

An alternative approach was taken by the CERTS project in the United States (Figure 7.11).

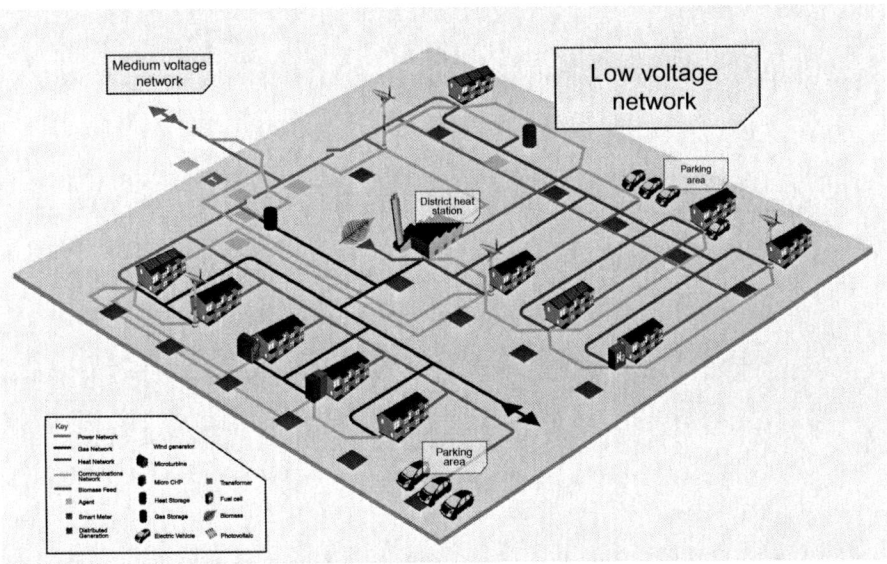

Figure 7.8   Illustration of a multi-energy vector microgrid

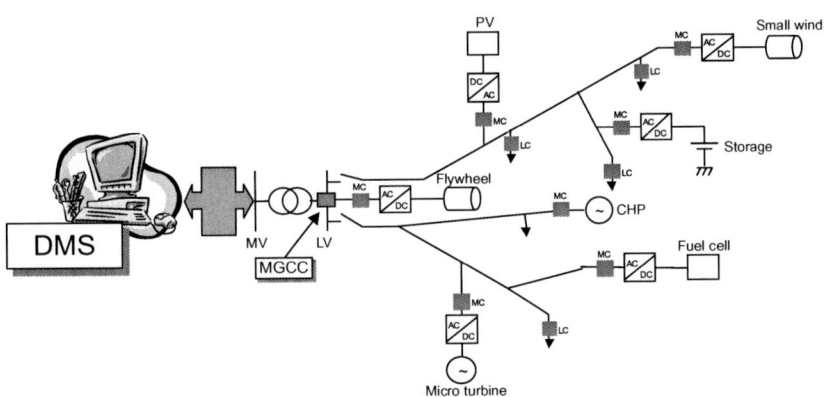

Figure 7.9   European Union MicroGrid Concept [18]

Battery energy storage was provided on the DC bus of each microgenerator. All the microgenerators had a plug-and-play capability and the feeders that supplied sensitive loads could be isolated from the distribution network by opening the static switch.

Several demonstration microgrid projects have been undertaken in Japan. The main objective was to balance fluctuations in load and intermittent generation so that the microgrid appeared benign to the power system. In the Sendai project (Figure 7.12) both DC and AC loads are supplied and a dynamic voltage restorer was used to improve the power quality to high value loads.

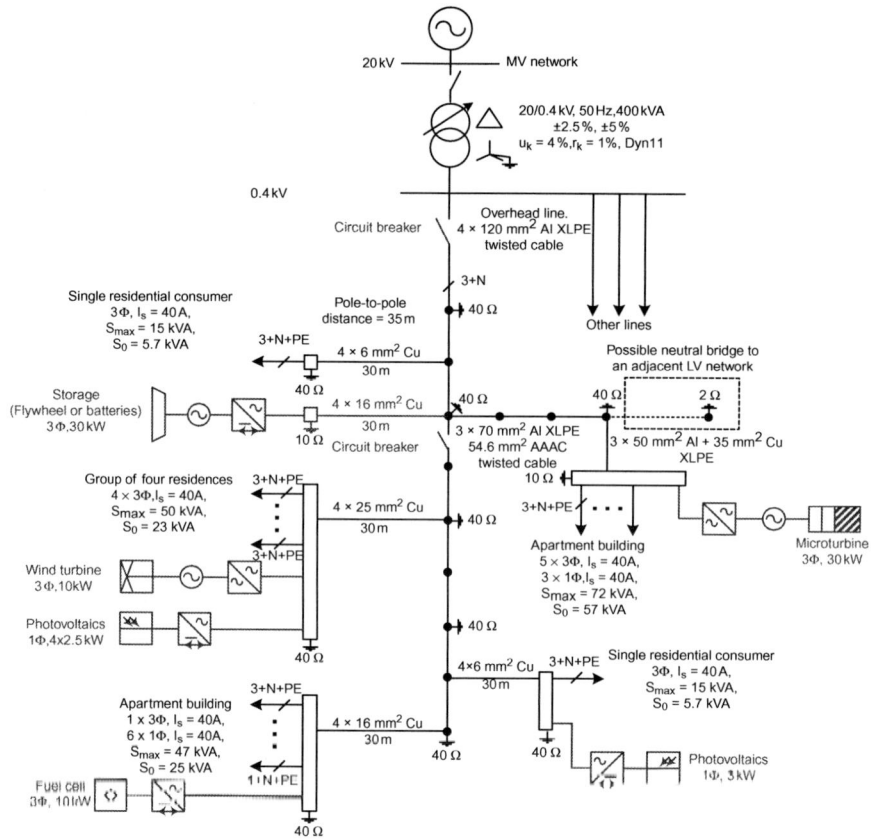

Figure 7.10   European Union MicroGrid Test System [18]

7.4.2   Microgrid control

A microgrid can operate in one of two states: grid-connected and islanded.

When the microgrid is grid connected, it receives its frequency and voltage from the distribution network at the point of connection. The microgenerators can then be described as 'grid takers' and function as sources of real and reactive power. The converters of the microgenerators measure the system frequency/phase using a phase locked loop and inject appropriate currents.

During grid-connected operation, a microgrid can be controlled as a 'good citizen' or as a 'model citizen' [19]. A good citizen approach requires the microgrid to comply with distribution network rules without participating in the operation of the main power system. Its attributes include reducing the frequency and extent of electrical power shortages, improving local voltage and not causing deterioration of the power quality supplied to the customers. A model citizen approach provides ancillary services to the main power system. Thus, a microgrid acting as a model citizen becomes an important component sustaining the stable operation of the power system.

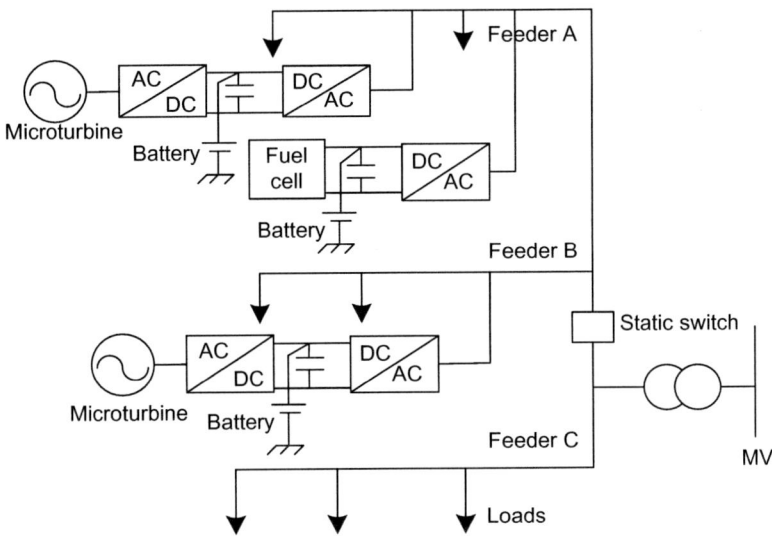

Figure 7.11   CERTS MicroGrid Topology [19, 21]

During islanded operation, a microgrid establishes its own voltage and frequency and maintains acceptable power quality. One or more of the microgenerators are then 'grid formers'. In a traditional power system, frequency is controlled by the large central power plants and the transmission grid voltage is regulated by transformer tap changers and reactive power from these generators. It is difficult to keep the islanded microgrid stable due to the varying loads and intermittent output of the

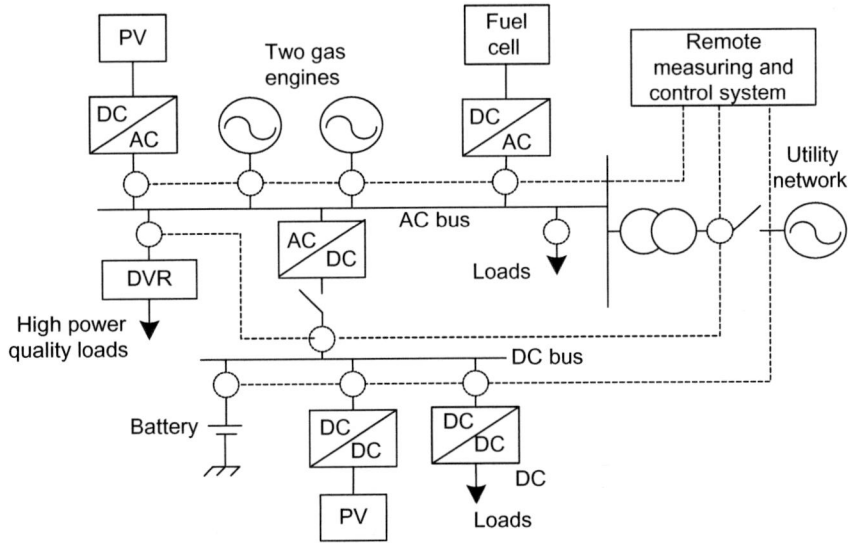

Figure 7.12   Sendai Microgrid Topology [19]

microgenerators. The frequency is regulated by controlling the output power of the microgenerators, by load shedding (demand side management) and using the energy store(s). Local voltage is influenced by both active and reactive power of the micro sources due to the high R/X ratio in the microgrid.

Stable transition between grid-connected and islanded mode is a key consideration. If the microgrid is absorbing or supplying power to the main grid before disconnection, a power imbalance occurs in the transition from grid-connected to islanded mode. When a microgrid is reconnected to the main grid from the islanded mode, it is synchronised in the conventional manner by ensuring the magnitude and phase of the voltages across the synchronising device are equal. A high-speed static switch with appropriate sensing capability may be used for disconnection and synchronisation.

7.4.3   Control strategies during islanded operation [22]

Two control strategies of the microgenerators are commonly used, peer–peer and master–slave control.

In peer–peer control, each microsource has an equivalent status and no single component, such as the master controller or the central energy storage unit, is critical for the operation of the microgrid. The microgrid can continue operating with the connection or disconnection of any microgenerator, if the energy requirements are still satisfied, and thus plug-and-play capability is achieved.

The droop characteristics, shown in Figure 7.13, may be used to control the microgenerator, but unlike in a large power system in some microgrids the output real and reactive powers of the microgenerators are measured and used to set their frequency and voltage.

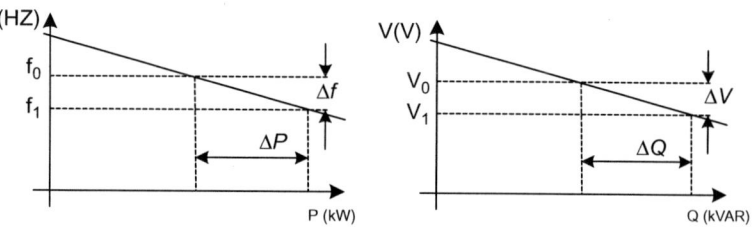

Figure 7.13   Microgrid droop characteristics

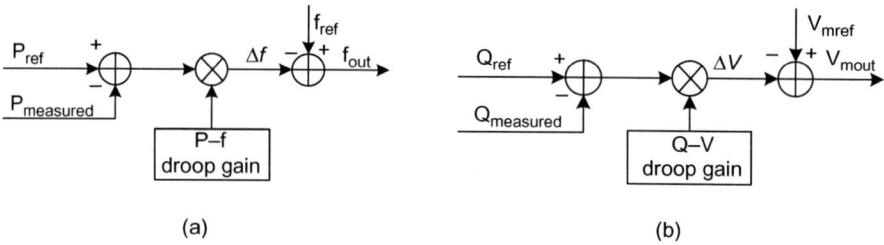

Figure 7.14   P–f and Q–V droop control

A reverse droop control method (P–V and Q–f) has also been applied because of the resistive nature of the microgrid.

In master–slave control the slave modules receive instructions from the master through communication channels. Two master–slave control modes have been investigated. One used a single microsource as the master, and the other used a central controller to supervise microsources and loads. The central controller concept has been widely used in microgrid demonstration projects.

A flywheel storage unit has been used as the master to regulate the frequency and voltage of the islanded microgrid, and the other micro sources were slaves whose outputs were operated under PQ control. The master unit used f–P and V–Q droop control as shown in Figure 7.14. An energy storage unit with large capacity was required in this scheme.

A central controller has been used for steady-state microgrid control with dynamic control delegated to the microsources. A microgrid central controller was used to coordinate the set points of the microsource controllers. These set points were then used by the individual microsource control systems. The frequency and voltage were primarily regulated by the microsource master, and the microgrid central controller changed the set points if the frequency and voltage exceeded their limits.

The various research and demonstration projects to develop microgrids have provided much valuable information and experience. However, they have also highlighted the advantages of an interconnected power system. Under present commercial conditions, it is only in unusual circumstances that the benefits from islanded operation of a small, low voltage microgrid are likely to justify the complexity and costs.

References

1. Dondi P., et al. 'Network integration of distributed power generation'. *Journal of Power Sources*. 2002;**106**:1–9.
2. Djapic P., et al. 'Taking an active approach'. *IEEE Power and Energy Magazine*. July–Aug. 2007;**5**(4):68–77.
3. Pecas Lopes J.A., et al. 'Integrating distributed generation into electric power systems: A review of drivers, challenges and opportunities'. *Electric Power Systems Research*. 2007;**77**(9):1189–1203.
4. Bayod-Rujula A.A. 'Future development of the electricity systems with distributed generation'. *Energy*. 2009;**34**:377–383.
5. Bollen M.H.J., et al. 'The active use of distributed generation in network planning'. *CIRED 20th International Conference on Electricity Distribution*; Prague, 8–11 June 2009, Paper 0150.
6. Hughes F.M., Anaya-Lara O., Jenkins N., Strbac G. 'Control of DFIG-based wind generation for power network support'. *IEEE Transaction on Power Systems*. 2005;**20**(4):1958–1966.

7. Borbely A.M. (ed.). *Distributed Generation: The Power Paradigm for the New Millennium.* CRC Press; 2001.
8. D'adamo C., Jupe S., Abbey C. 'Global survey on planning and operation of active distribution networks – Update of CIGRE C6.11 working group activities'. *Electricity Distribution, 2009, 20th International Conference and Exhibition*; 8–11 June 2009, Paper 0555.
9. Lund P. 'The Danish Cell Project – Part 1: Background and general approach'. *IEEE Power Engineering Society General Meeting*; 2007, Florida.
10. Cherian S., Knazkins V. 'The Danish Cell project – Part 2: Verification of control approach via modeling and laboratory tests'. *IEEE Power Engineering Society General Meeting*; 2007, Florida.
11. Hird M., Helder L., Li H., Jenkins N. 'Network voltage controller for distributed generation'. *IEE Proceeding Generation, Transmission and Distribution.* 2004;**151**(2):150–156.
12. Thornley V., Hill J., Lang P., Reid D. 'Active network management of voltage leading to increased generation and improved network utilisation'. *IET-CIRED Seminar SmartGrids for Distribution*; 23–24 June 2008, Frankfurt.
13. Taylor P., et al. 'Integrating voltage control and power flow management in AuRA-NMS'. *IET-CIRED Seminar SmartGrids for Distribution*; 23–24 June 2008, Frankfurt.
14. Pudjianto D., Ramsay C., Strbac G. 'Virtual power plant and system integration of distributed energy resources'. *Renewable Power Generation, IET.* 2007;**1**(1):10–16.
15. Barnes M., et al. 'Real-World MicroGrids an overview'. *System of Systems Engineering, 2007. SoSE'07. IEEE International Conference*; 16–18 April 2007, pp. 1–8, San Antonio, TX.
16. Hatziargyriou N., Asano H., Iravani R., Marnay C. 'Microgrids: An overview of ongoing research, development, and demonstration projects'. *IEEE Power and Energy Magazine.* August 2007, pp. 78–94.
17. Barnes M., Dimeas A., Engler A. 'MicroGrid laboratory facilities'. *International Conference on Future Power Systems*; November 2005, pp. 1–6.
18. European Research Project More MicroGrids. [Online]. Available from http://www.microgrids.eu/default.php
19. Xiao Z., Wu J., Jenkins N. 'An overview of microgrid control.' *Intelligent Automation and Soft Computing.* 2010;**16**(2):199–212, ISSN 1079-8587.
20. Lasseter R.H. 'MicroGrids and distributed generation.' *Journal of Energy Engineering American Society of Civil Engineers.* 2007;**133**(3):144–149.
21. Nikkhajoei H., Lasseter R.H. 'Distributed generation interface to the CERTS microgrid.' *IEEE Transactions on Power Delivery.* 2009;**24**(3):1598–1608.
22. Peças Lopes J.A., Moreira C.L. 'Defining control strategies for MicroGrids islanded operation.' *IEEE Transactions on Power Systems.* 2006;**21**(2):916–924.

Tutorial I
AC electrical systems

Aylesford Co-generation Plant
The plant produces 220 MW heat and 98 MW of electrical energy [National Power PLC]

I.1 Introduction

In all public electricity supply systems, the mains voltage alternates at 50 or 60 cycles per second (Hz) and when a load is connected it draws an alternating current.

An alternating current or voltage changes its polarity periodically to give a wave-like shape on an oscilloscope trace. Sine, square and even saw-tooth waves are used in different electronic circuits, but this chapter considers only the sine wave, which is the shape of the AC mains.

A generator produces a three-phase alternating voltage and this voltage is increased (stepped-up) for long distance transmission (Figure 1.1, shown on page 2). The transmission voltages are then stepped-down and the power distributed to loads. Low voltage final distribution circuits typically use four wires (in Europe, three phase wires each at 230 V and a neutral wire that provides the zero volt reference). Single-phase loads (e.g. houses) are connected across two wires (to one phase wire and the neutral wire) and three-phase loads (industry and commercial buildings) are connected to all four wires.

I.2   Alternating current (AC)

A periodic voltage or current waveform is described by its:

1. Period (T)
    The time taken for one cycle of voltage, or current, is known as the period, with the symbol T and is measured in seconds. The period can be measured from any point on one cycle to the corresponding point on the next (Figure I.1).
2. Frequency (f)
    The number of cycles that a waveform completes in 1s is its frequency. Frequency has the symbol f and is measured in Hertz (Hz). Period and frequency are related by: $f = 1/T$ Hz; i.e. a longer period results in a lower frequency, while a shorter period gives a higher frequency.
3. Peak value (V_m)
    The positive or negative maximum value is known as the peak value. The peak value is measured in volts or amperes.
4. Peak-to-peak value ($2V_m$)
    The magnitude between the negative peak and positive peak is called the peak-to-peak value.

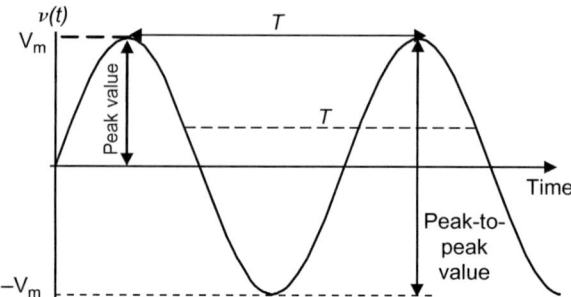

Figure I.1   Sinusoidal voltage waveform and definitions

I.3   Root mean square value of voltage and current

One measure of a sinusoidal voltage or current is obtained by considering the average power dissipated in a resistor.

Applying Ohm's law to the two circuits shown in Figure I.2:

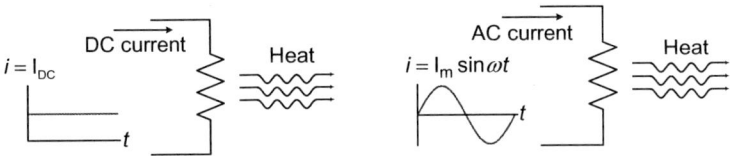

Figure I.2   Resistor fed by DC and AC

For DC: $V_{DC} = I_{DC} \times R$. Therefore, the power dissipated as heat is: $P = V_{DC} \times I_{DC} = I_{DC}^2 \times R$.

For AC: If the current through the resistor is $i(t) = I_m \sin(\omega t)$, then $v(t) = I_m R \sin(\omega t)$.

The instantaneous power dissipation is: $p_{int}(t) = v(t) \times i(t) = I_m^2 R \sin^2(\omega t)$.

The average power dissipation: $P_{ave} = 1/T \int_0^T I_m^2 R \sin^2(\omega t) dt$.

Substituting $\sin^2(\omega t) = (1 - \cos(2\omega t))/2$, the following equation is obtained:

$$P_{ave} = \frac{1}{T} \int_0^T I_m^2 R \left[\frac{1 - \cos(2\omega t)}{2} \right] dt$$

$$= \frac{1}{2T} \int_0^T I_m^2 R \times dt - \frac{1}{2T} \int_0^T I_m^2 R \cos(2\omega t) dt = \frac{I_m^2 R}{2} \tag{I.1}$$

The average heating effect in both circuits is equal, if

$$I_{DC}^2 \times R = \frac{I_m^2 R}{2}$$

$$I_{DC} = \frac{I_m}{\sqrt{2}} \tag{I.2}$$

The value of the AC current, that gives the same heating effect as a DC current into a resistor, is termed the root mean square (rms) current and has the value $I_m/\sqrt{2}$ for a sine wave. Alternating voltages and currents are usually expressed by their rms values.

I.4   Phasor representation of AC quantities

Consider a vector of length V_m rotating anti-clockwise at angular velocity ω (Figure I.3). As the vector **OA** rotates, its projection on the y axis will describe a sinusoidal signal (V_a). Similarly when the vector **OB** rotates, its projection on the y axis describes V_b. The length of the vector, **OA** or **OB**, corresponds to the peak value of the sinusoidal signal (V_m) and the angle between the vectors **OA** and **OB** corresponds to the angle between any identical points in the two waveforms.

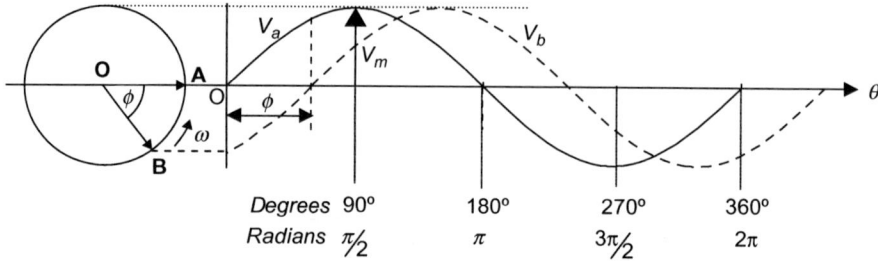

Figure I.3   Development of a sinusoidal waveform by revolving a phasor

If the rotation of vectors **OA** and **OB** are at ω rad/s, then the angle θ (from the origin) of the sinusoidal waveform V_a at time t is equal to ωt rad. Therefore, the equation for the sinusoidal signal V_a is $v_a(t) = V_m \sin \theta = V_m \sin(\omega t)$. Similarly the equation for the sinusoidal signal V_b is $v_b(t) = V_m \sin(\omega t + \phi)$.

From Figure I.3: The period of the signal V_a is $2\pi/\omega$.

If the frequency of the signal is f, then: $1/f = 2\pi/\omega$; i.e. $\omega = 2\pi f$.

The vectors **OA** and **OB** which rotate at the same angular velocity as the sinusoidal signal and which represent the magnitude and the angle of the sinusoidal signals are called *phasors*. The angle between two phasors describes how far ahead or behind a point on one sinusoidal signal is to the same point of the other signal and is called the phase angle. If the phasors of two signals coincide with each other, then the two signals are in phase. If one is ahead of the other signal by an angle ϕ (measured anti-clockwise) then it *leads* the second one by angle ϕ. If one is behind the other signal by an angle ϕ (measured clockwise) then it *lags* the second one by angle ϕ.

Two phasors (**OA** and **OB**) which represent signals V_a and V_b can be written in two forms as given in Table I.1. Even though the lengths of phasors **OA** and **OB** in Figure I.3 are equal to the peak value of sinusoidal signal, in phasor representation, rms values are more usually used.

Table I.1   Time and phase representation of AC quantities

Signal	Time representation	Phasor representation
V_a	$V_a = V_m \sin(\omega t + 0)$	$\mathbf{V_a} = \dfrac{V_m}{\sqrt{2}} \angle 0$
V_b	$V_b = V_m \sin(\omega t - \phi)$	$\mathbf{V_b} = \dfrac{V_m}{\sqrt{2}} \angle -\phi$

The form $\mathbf{V}_\varphi = V\angle\phi$ is called the polar form of phasor $\mathbf{V}_\varphi$ and $\mathbf{V}_\varphi = V[\cos\phi + j\sin\phi]$ is called the rectangular or Cartesian form of phasor $\mathbf{V}_\varphi$.[1] In this book, phasors are given in bold ($\mathbf{V}_\varphi$) and their magnitude is given normal type (not bold) ($V = |\mathbf{V}_\varphi|$). The diagram representing phasor $\mathbf{V}_\varphi$ (by its magnitude V and angle ϕ) is shown in Figure I.4 (drawn using rms quantities) and is called the phasor diagram.

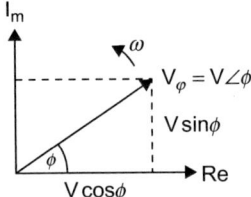

Figure I.4   Phasor diagram of $\mathbf{V}_\varphi$ (shown on real and imaginary axes)

I.5   Resistors, inductors and capacitors on AC circuits

I.5.1   Resistance in an AC circuit

Figure I.5 shows a resistor, R, connected across an AC voltage source $v(t) = V_m\sin(\omega t)$. The current through the resistor is determined by Ohm's law and is given by: $i(t) = V_m/R\sin(\omega t)$. If the voltage phasor is $\mathbf{V}$ and current phasor is $\mathbf{I}$, then $\mathbf{V} = \mathbf{I}R$.

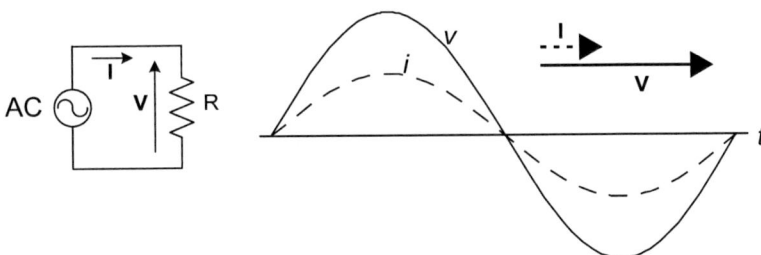

Figure I.5   Time domain and phasor representation of voltage and current in R

I.5.2   Inductance in an AC circuit

The relationship between current and voltage for a pure inductance is $v = L(di/dt)$. For an applied voltage v as shown in Figure I.6, the current can be obtained by

---

[1] j is the imaginary operator ($\sqrt{-1}$) and in polar form $j = 1\angle 90°$.

integrating the voltage:

$$i(t) = \frac{1}{L}\int_0^t v\,dt = \frac{1}{L}\int_0^t V_m \sin(\omega t)\,dt = -\frac{V_m}{\omega L}\cos \omega t$$

$$= \frac{V_m}{\omega L}\sin(\omega t - \pi/2)$$

The voltage and current waveforms for the inductive circuit are shown in Figure I.6. The current through the inductor lags the voltage by 90°. If the voltage phasor is **V** and current phasor is **I**, then $\mathbf{V}/\mathbf{I} = \omega L \angle 90° = j\omega L$. The quantity $j\omega L$ is the impedance of the inductor and has the unit of ohms. The magnitude of the impedance, i.e. $X_L = \omega L$, is referred to as inductive reactance, also with units of ohms. When the inductor is represented by its reactance, then Ohm's law can be applied to the phasors of voltage and current.

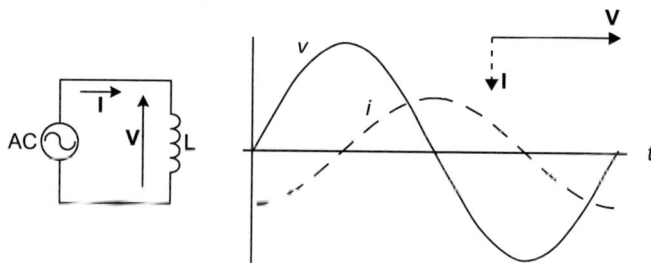

Figure I.6   Time domain and phasor representation of voltage and current in L

I.5.3   Capacitance in an AC circuit

The relationship between current and voltage for a pure capacitance is $i = C(dv/dt)$. As shown in Figure I.7, if a sinusoidal voltage v is applied across the capacitor, the current through the capacitor is given by:

$$i(t) = C\frac{d}{dt}(V_m \sin \omega t)$$

$$= \omega C V_m \cos \omega t \qquad\qquad (I.4)$$

$$= \omega C V_m \sin(\omega t + (\pi/2))$$

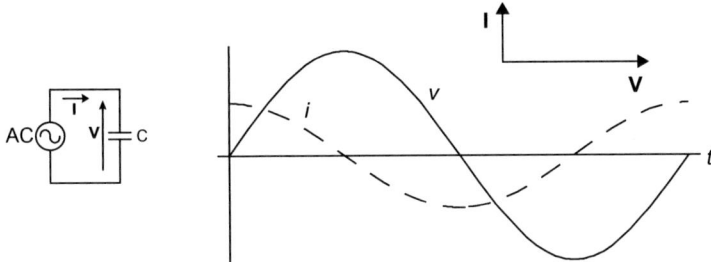

Figure I.7   Time domain and phasor representation of voltage and current in C

If the voltage phasor is $\mathbf{V}$ and the current phasor is $\mathbf{I}$, $\mathbf{V}/\mathbf{I} = (1/\omega C)\angle -90° = 1/j\omega C$. The quantity $1/j\omega C$ is the impedance of the capacitor and has the unit of ohms. The magnitude of the impedance, i.e. $X_C = 1/\omega C$, is referred to as the capacitive reactance, again with the unit of ohms. When the capacitor is represented by its reactance, then Ohm's law can be applied to the phasors of voltage and current.

I.5.4   R–L in an AC circuit

A series R–L circuit is shown in Figure I.8. The current through the circuit is limited by the impedance of the circuit, given by $\mathbf{Z} = R + jX_L$. The phase angle between the applied voltage and current is given by $\tan^{-1}(\omega L/R)$.

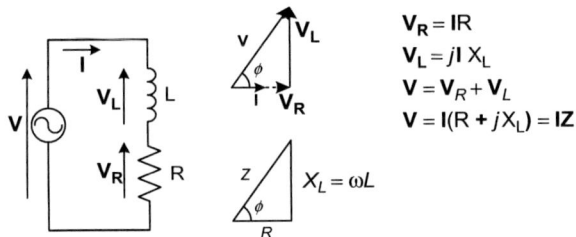

Figure I.8   Series R–L circuit, phasor and impedance diagrams

---

EXAMPLE I.1

Calculate the resistance and inductance or capacitance in series for each of the following impedances, assuming the frequency to be 60 Hz:

(a) $50 + j30 \; \Omega$

(b) $40\angle -60° \Omega$

Answer:

(a) $50 + j30 \; \Omega = R + j\omega L$
       Comparing: $R = 50 \, \Omega$ and $\omega L = 30 \, \Omega$
       When the frequency is 60 Hz: $\omega = 2\pi f = 2\pi \times 60 = 377 \, \text{rad/s}$
       Therefore, $L = 30/377 = 79.6 \times 10^{-3} \, \text{H} = 79.6 \, \text{mH}$

(b) $40\angle - 60° = 20 - j34.64\,\Omega = R + (1/j\omega C) = R - (j/\omega C)$
Comparing: $R = 20\,\Omega$ and $1/\omega C = 34.64\,\Omega$
Since $\omega = 377\,\text{rad/s}$, $C = 1/(377 \times 34.64) = 76.6 \times 10^{-6}\,\text{F} = 76.6\,\mu\text{F}$

EXAMPLE I.2

A voltage $v(t) = 170\sin(377t)$ volts is applied across a winding having a resistance of $2\,\Omega$ and an inductance of $0.01\,\text{H}$. Write down an expression for the rms values of the voltage and current phasors in rectangular notation. Draw the phasor diagram.

Answer:
$R = 2\,\Omega$ and $L = 0.01\,\text{H}$
Since the voltage is $v(t) = 170\sin(377t)$: $\omega = 377\,\text{rad/s}$
Voltage as a phasor $= (170/\sqrt{2})\angle 0° = 120.2\angle 0° = 120.2 + j0\,\text{V}$
$\omega L = 377 \times 0.01 = 3.77\,\Omega$
Impedance of the winding $= 2 + j3.77\,\Omega = 4.27\angle 62.05°\,\Omega$
Current as a phasor $= 120.2\angle 0°/4.27\angle 62.05° = 28.13\angle - 62.05° = 13.18 - j24.85\,\text{A}$

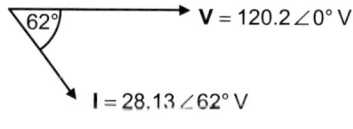

$\mathbf{V} = 120.2\angle 0°\,\text{V}$

$62°$

$\mathbf{I} = 28.13\angle 62°\,\text{V}$

Phasor diagram

EXAMPLE I.3

An impedance of $2 + j6\,\Omega$ is connected in series with two impedances of $10 + j4$ Ω and $12 - j8\,\Omega$, which are in parallel. Calculate the magnitude and the phase angle of the main current when the combined circuit is supplied at 200 V.

Answer:

$\mathbf{Z_t}$

$\mathbf{Z_2}$

$\mathbf{Z_1}$

$\mathbf{Z_3}$

$\mathbf{Z_p}$ 200 V

$\mathbf{Z_1} = 2 + j6\,\Omega$

$\mathbf{Z_2} = 10 + j4\,\Omega$

$\mathbf{Z_3} = 12 - j8\,\Omega$

$$\mathbf{Z_p} = \frac{\mathbf{Z_2 Z_3}}{\mathbf{Z_2 + Z_3}} = \frac{[10 + j4][12 - j8]}{10 + j4 + 12 - j8} = 6.95 - j0.19$$

$$\mathbf{Z_t} = \mathbf{Z_1} + \mathbf{Z_p} = 8.95 + j5.81 = 10.67\angle 33°$$

$$\mathbf{I} = \frac{200}{10.67\angle 33°} = 18.74\angle -33°$$

Magnitude of the current = 18.74 A
Angle of the current = 33° and lags the voltage

I.6   Power in AC circuits

Instantaneous power in an AC circuit is the product of instantaneous voltage and current.

For a pure resistor $v(t) = R \times i(t)$ and the instantaneous power, $p(t)$, is:

$$p(t) = v(t) \times i(t) = v(t)^2/R = i(t)^2 \times R \tag{I.5}$$

If $v(t) = V_m \sin \omega t$, the average power dissipation in the resistor is given by:

$$P_{ave} = \frac{1}{T}\int_0^T p(t)dt = \frac{V_m^2}{R \times T}\int_0^T \sin^2 \omega t \times dt = \frac{V_m^2}{R \times T}\int_0^T \left[\frac{1 - \cos(2\omega t)}{2}\right] \times dt$$

$$= \frac{V_m^2}{2R} = \frac{V^2}{R} = I^2 R \tag{I.6}$$

where V and I are rms quantities.

For a pure inductor $v(t) = L(di(t)/dt)$ and the instantaneous power is:

$$p(t) = v(t) \times i(t) = L \times i(t) \times \frac{d}{dt}i(t) \tag{I.7}$$

If $i(t) = I_m \sin \omega t$, the average power dissipation in the inductor is given by:

$$P_{ave} = \frac{\omega L I_m^2}{T}\int_0^T \sin \omega t \times \cos \omega t \times dt = 0 \tag{I.8}$$

Figure I.9 shows the instantaneous power associated with two circuits one having a pure resistor and the other having a pure inductor. In the circuit having a resistor, the instantaneous power varies at double the frequency of the voltage and current with a positive average value. The power of this circuit converts electrical energy into heat and is called active power. In the circuit having an inductor, the instantaneous power alternates with a zero average value. The power associated with energy oscillating in and out of an inductor is called reactive power. The active power is measured in watt or kilowatt, and the reactive power is measured in Var or kiloVar.

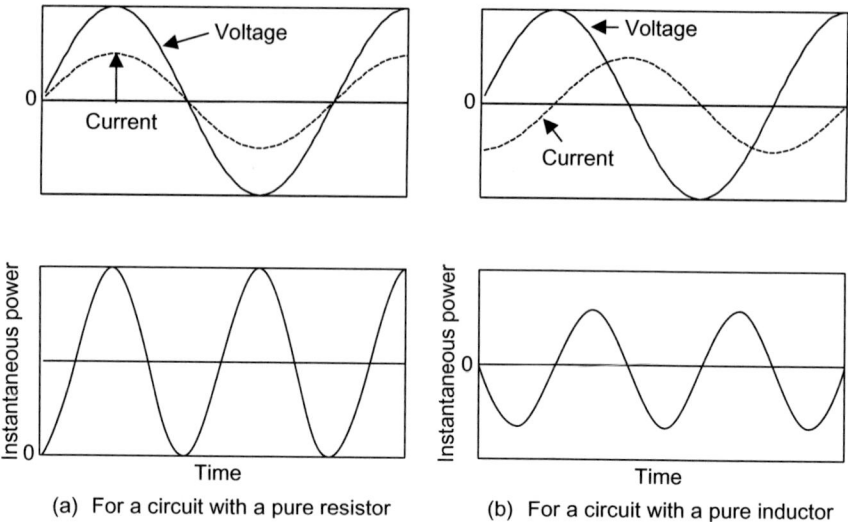

(a) For a circuit with a pure resistor   (b) For a circuit with a pure inductor

Figure I.9   Instantaneous power for a circuit with a resistor and an inductor

Now consider a resistor and an inductor connected across an AC voltage source as shown in Figure I.10(a). The phasor diagram is shown in Figure I.10(b). The instantaneous current can now be divided into two components: a component in phase with the voltage, I_P (which causes the active power), and a component having 90° phase shift, I_Q (which causes the reactive power). The instantaneous power associated with this circuit and its active and reactive power components are shown in Figure I.11.

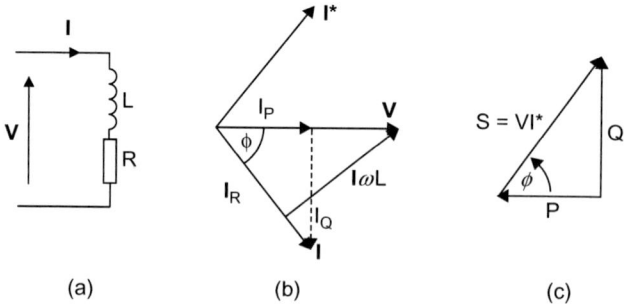

(a)   (b)   (c)

Figure I.10   An R–L circuit (inductive load)

From Figure I.10(c):

Active power, $P = V \times I_P = VI \cos \phi$ (W)   (I.9)

Reactive power, $Q = V \times I_Q = VI \sin \phi$ (VAr)   (I.10)

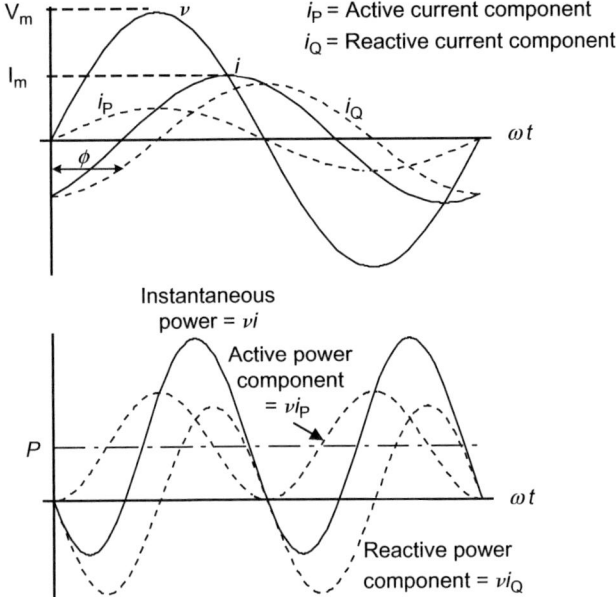

Figure I.11   Active and reactive power associated with an inductive load

It is conventional to define **S**, the apparent power, as **VI*** (see Figure I.10(b) where I* is shown). In Figure I.10(c), $S = \sqrt{P^2 + Q^2} = VI^*$. S is measured in volt-ampere or kilovolt-ampere. The cosine of the angle ϕ (cos ϕ) is called the power factor of the circuit.

Most industrial and large commercial electricity consumers have pre-dominantly inductive loads. For a given applied voltage and real power load, the current drawn is high if the power factor is low. This in turn increases the size of the distribution cables required both within the consumer premises and on the utility side, the size of the transformers and losses in cables and transformers.

EXAMPLE I.4

A single-phase motor of 10 kW operates at a power factor of 0.8 lagging when connected to a 230 V, 50 Hz supply. It is proposed to improve the power factor to 0.9 lagging by connecting a capacitor across the load. Calculate the kVAr rating of the capacitor.

Answer:

As shown in Figure I.7, for a capacitor $\phi = 90°$ and therefore from (I.9) and (I.10), $P = 0$ and $Q = VI$; that is a capacitor acts as a source of reactive power. Therefore, if a capacitor is connected across the motor, a part of the reactive power drawn by the motor is supplied by the capacitor, thus reducing reactive power drawn from the AC mains. This is called power factor correction.

Power factor angle before correcting the power factor $= \cos^{-1} 0.8 = 36.87°$
Power factor angle after correcting the power factor $= \cos^{-1} 0.9 = 25.84°$

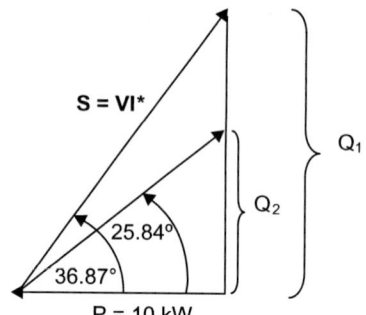

Reactive power drawn from the supply before power factor correction[2]

$$Q_1 = 10 \tan 36.87°$$
$$= 7.5 \, \text{kVAr}$$

Reactive power drawn from the supply after power factor correction

$$Q_2 = 10 \tan 25.84°$$
$$= 4.84 \, \text{kVAr}$$

That is a capacitor connected across the motor should supply the reactive power of $(Q_1 - Q_2)$ locally. Thus, the kVAr rating of the capacitor

$$= Q_1 - Q_2$$
$$= 7.5 - 4.84 = 2.66 \, \text{kVAr}$$

I.7   Generation of three-phase voltages

If three coils, shown in Figure I.12(a), which are physically displaced 120° are supplied with a three-phase AC voltage, then the resultant voltage in each coil is displaced in time by 120° as shown in Figure I.12(b).

In a three-phase system, the voltage in B phase lags the voltage in A phase by 120° $(2\pi/3 \, \text{rad})$ and the voltage in C phase lags the voltage in A phase by 240° $(4\pi/3 \, \text{rad})$. The three-phase system can be represented by the phasor diagram of Figure I.13. The corresponding mathematical description of the three-phase system is also given in that figure.

---

[2] From (I.9) and (I.10), $Q = p \tan \phi$.

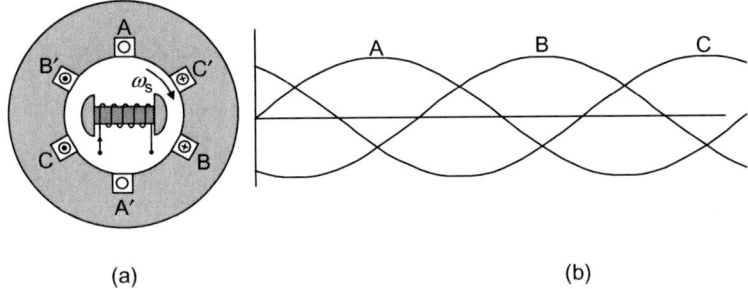

Figure I.12   Three-phase voltages

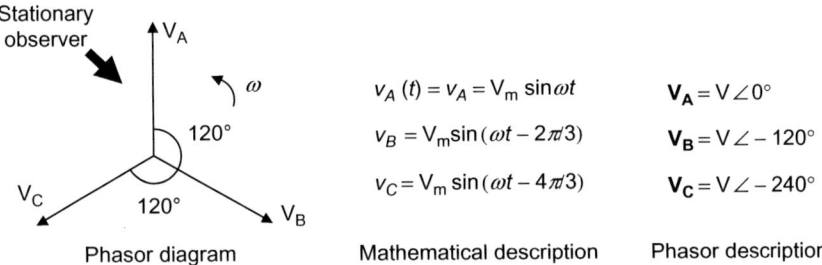

| Phasor diagram | Mathematical description | Phasor description |

Figure I.13   Different descriptions of the three-phase system (Instantaneous values are given in the form of v (not in the form v(t)) in the subsequent sections)

A stationary observer sees the phasor diagram in Figure I.13 will see coils A, B and C passing him in time in that order. Therefore, the phase sequence of the supply is ABC.

I.8   Connection of three-phase windings

The windings in the three phases of Figure I.12(a) are always connected together so that three-phase voltages are carried on three or four wires. Three windings can be connected in star (wye) or delta.

I.8.1   Star connection

A star connection can be obtained by connecting terminals A', B' and C' (see Figure I.12(a)) together to form a neutral point (N) as shown in Figure I.14. From this connection, three or four wires can be taken to the load thus forming a three-phase, star-connected, three-wire or four-wire system. Two voltages can be defined: the phase voltage (the voltage between the neutral and any terminal, V_{AN}, V_{BN} and V_{CN}) and the line voltage (the voltage between any two terminals, V_{AB}, V_{BC} and V_{CA}). In the star connection, the phase current and line current are the same and denoted by I_A, I_B and I_C in Figure I.14.

196   *Distributed generation*

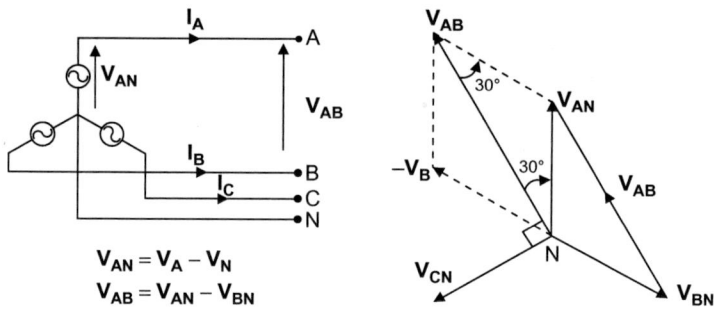

Figure I.14   Star-connected three-phase source

The phasor diagram in Figure I.14 shows the line voltage V_{AB}. Assuming that the rms value of the magnitude of phase voltages V_{AN}, V_{BN} and V_{CN} is V_p and the rms value of the magnitude of line voltages V_{AB}, V_{BC}, V_{CA} is V_L, then from the phasor diagram:

$$V_{AB} = |V_{AN}| \times \cos 30° + |V_{BN}| \times \cos 30°$$
$$V_L = 2 \times V_p \times \cos 30° \quad\quad\quad\quad (I.11)$$
$$V_L = \sqrt{3} V_p$$

For example, if the phase voltage is 230 V, then the line voltage is $\sqrt{3} \times 230 = 400V$.

I.8.2   Delta connection

By connecting terminal A′ to B, B′ to C and C′ to A, the delta connection can be formed as shown in Figure I.15. In the delta connection, there is no neutral, and line voltages and phase voltages are the same. If $V_{AB} + V_{BC} + V_{CA} = 0$, there is no circulating current in the delta loop.

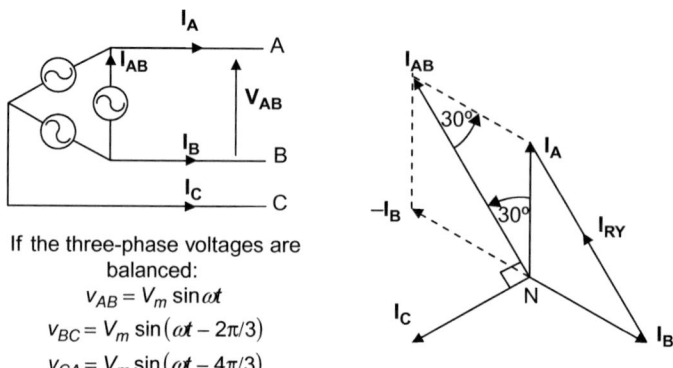

If the three-phase voltages are
balanced:
$v_{AB} = V_m \sin \omega t$
$v_{BC} = V_m \sin(\omega t - 2\pi/3)$
$v_{CA} = V_m \sin(\omega t - 4\pi/3)$

Figure I.15   Delta-connected three-phase system

The relationship between the line and phase currents can be obtained by assuming the three phase currents are balanced (i.e. the connected loads are equal). From Figure I.15, the magnitude of the line current is equal to $\sqrt{3}$ times the magnitude of the phase current.

I.9 Connection of loads

Loads can also be connected in star or delta. If the loading on each phase is equal, then that load is called a balanced three-phase load. Figure I.16(a) and (b) shows balanced star-connected and delta-connected loads.

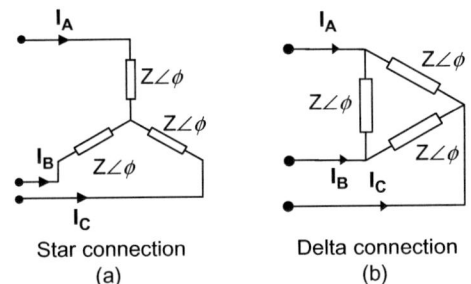

Star connection
(a)

Delta connection
(b)

Figure I.16 Star and delta connection of loads

I.10 Three-phase four-wire system

A three-phase star-connected generation system is connected to a three-phase star-connected load as shown in Figure I.17 to form a three-phase four-wire system.

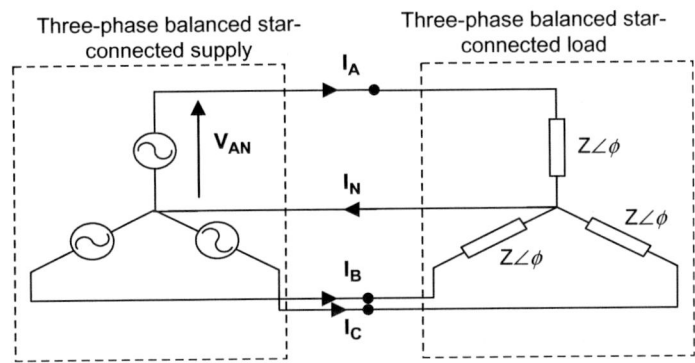

Three-phase balanced star-connected supply

Three-phase balanced star-connected load

Figure I.17 Configuration of the three-phase four-wire system

When the load is balanced, the current in the neutral wire is zero.

$$
\begin{aligned}
\mathbf{I_N} &= \mathbf{I_A} + \mathbf{I_B} + \mathbf{I_C} \\
&= \frac{\mathbf{V_{AN}}}{Z\angle\phi} + \frac{\mathbf{V_{BN}}}{Z\angle\phi} + \frac{\mathbf{V_{CN}}}{Z\angle\phi} \\
&= \frac{V_m}{Z\angle\phi}[\sin\omega t + \sin(\omega t - 2\pi/3) + \sin(\omega t - 4\pi/3)] = 0
\end{aligned}
\tag{I.12}
$$

However, when the loads are not balanced (i.e. the three loads are not equal), then there will be a current flowing through the neutral wire.

I.11   Three-phase delta-connected three-wire system

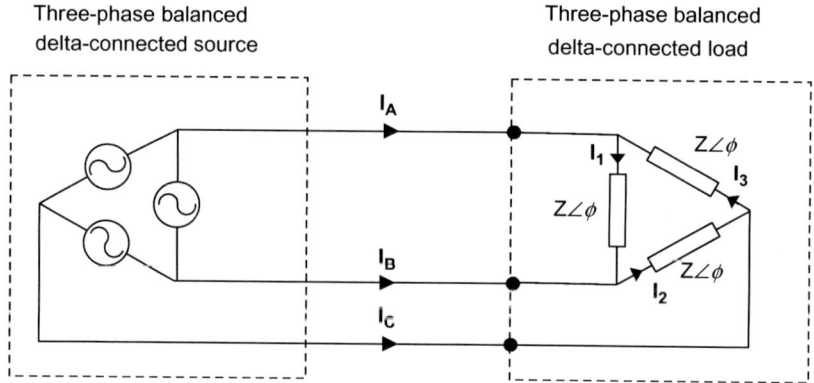

Figure I.18   Three-phase delta-connected source and a load

A three-phase delta-connected generator is connected to a three-phase delta-connected load as shown in Figure I.18 to form the three-phase, three-wire system. In the figure $\mathbf{I_A}$, $\mathbf{I_B}$ and $\mathbf{I_C}$ are line currents, and $\mathbf{I_1}$, $\mathbf{I_2}$ and $\mathbf{I_3}$ are phase currents.

$$
\mathbf{I_A} = \mathbf{I_1} - \mathbf{I_3}, \ \mathbf{I_B} = \mathbf{I_2} - \mathbf{I_1} \text{ and } \mathbf{I_C} = \mathbf{I_3} - \mathbf{I_2}
\tag{I.13}
$$

The corresponding phasor diagram of the system is shown in Figure I.19. Defining the rms value of the magnitude of phase currents as I_p and the rms value of the magnitude of line currents as I_L, and then from the phasor diagram, we get:

$$
\begin{aligned}
I_L &= 2 \times I_p \times \cos 30° \\
I_L &= \sqrt{3} I_p
\end{aligned}
\tag{I.14}
$$

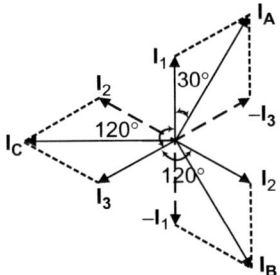

Figure I.19   Phasor diagram of currents in a delta-connected load

I.12   Power in three-phase system

The power in a three-phase load is the sum of the power consumed in each phase. Consider the load shown in Figure I.17.

For the load connected to the A phase:

$$v_{AN} = V_m \sin \omega t$$

$$i_A = \frac{V_m \sin \omega t}{Z \angle \phi} = I_m \sin(\omega t - \phi)$$

(I.15)

where $I_m = V_m/Z$.

For the other phases:

$$
\begin{aligned}
v_{BN} &= V_m \sin(\omega t - 2\pi/3) & v_{CN} &= V_m \sin(\omega t - 4\pi/3) \\
i_B &= I_m \sin(\omega t - 2\pi/3 - \phi) & i_C &= I_m \sin(\omega t - 4\pi/3 - \phi)
\end{aligned}
$$

(I.16)

Instantaneous power in A phase $= v_{AN} \times i_A = V_m \times I_m \times \sin \omega t \times \sin(\omega t - \phi)$

Instantaneous power in B phase $= v_{BN} \times i_B = V_m \times I_m \times \sin(\omega t - 2\pi/3) \times \sin(\omega t - 2\pi/3 - \phi)$

Instantaneous power in C phase $= v_{CN} \times i_C = V_m \times I_m \times \sin(\omega t - 4\pi/3) \times \sin(\omega t - 4\pi/3 - \phi)$

By adding these equations, the total power consumed by the three-phase load is obtained. It can be shown trigonometrically that

$$\sin \omega t \times \sin(\omega t - \phi) + \sin(\omega t - 2\pi/3) \times \sin(\omega t - 2\pi/3 - \phi)$$

$$+ \sin(\omega t - 4\pi/3) \times \sin(\omega t - 4\pi/3 - \phi) = \frac{3}{2} \cos \phi$$

Therefore, the total three-phase instantaneous power

$$= \frac{3}{2} V_m I_m \cos \phi = 3 \frac{V_m}{\sqrt{2}} \frac{I_m}{\sqrt{2}} \cos \phi = 3 V_p I_p \cos \phi$$

(I.17)

When (I.17) is compared with (I.9) for a single-phase case, it can be seen that the total three-phase instantaneous power is equal to the addition of active power in the three phases. Therefore, the active power, P, measured in watt for a three-phase circuit is given by:

$$P = 3V_p I_p \cos \phi \qquad (I.18)$$

where $\cos \phi$ is the power factor of the load.

Similarly as defined for a single-phase circuit, the apparent and reactive powers are defined for three-phase circuits.

$$\text{Apparent power} \quad S = 3V_p I_p \qquad (I.19)$$

$$\text{Reactive power} \quad Q = 3V_p I_p \sin \phi \qquad (I.20)$$

It is shown in Table I.2 that the power relationships for both star- and delta-connected loads are equal.

Table I.2   *Power relationships for star- and delta-connected loads*

Load	Star connected	Delta connected
Current and voltage relationship	$V_L = \sqrt{3}V_p$ and $I_L = I_p$	$V_L = V_p$ and $I_L = \sqrt{3}I_p$
Apparent power (VA)	$S = 3V_p I_p = 3\dfrac{V_L}{\sqrt{3}}I_L = \sqrt{3}V_L I_L$	$S = 3V_p I_p = 3V_L\dfrac{I_L}{\sqrt{3}} = \sqrt{3}V_L I_L$
Active power (kW)	$P = 3V_p I_p \cos \phi = \sqrt{3}V_L I_L \cos \phi$	
Reactive power (kVAr)	$Q = 3V_p I_p \sin \phi = \sqrt{3}V_L I_L \sin \phi$	

EXAMPLE I.5

A three-phase, wye-connected generator is driven from a water turbine that produces 15 kW mechanical shaft power. It is connected to a 500 V (V_L), three-phase, 60 Hz supply. The generator has an efficiency of 95% and operates at a power factor of 0.85 lagging (exporting VArs). Calculate

(a) The output apparent power and hence the current in the connecting circuit
(b) The active and reactive components of the current
(c) The reactive power

Answer:

(a) Output power = input power × efficiency = 15 × 0.95 = 14.25 kW
Apparent power, $S = \sqrt{3}V_L I_L$
Active power, $P = \sqrt{3}V_L I_L \cos \phi = S \cos \phi$

Therefore, output apparent power $= \dfrac{P}{\cos \phi} = \dfrac{14.25}{0.85} = 16.76\,\text{kVA}$

(b)   The line current, $I_L = \dfrac{P}{\sqrt{3}V_L\cos\phi} = \dfrac{14.25 \times 10^3}{\sqrt{3} \times 500 \times 0.85} = 19.36\,A$

Active component of the current $= I_L\cos\phi = 19.36 \times 0.85 = 16.45\,A$
Reactive component of the current $= I_L\sin\phi = 19.36 \times \sin(\cos^{-1}0.85) =$
10.2 A

(c)   Reactive power, $Q = \sqrt{3}V_L I_L\sin\phi = \sqrt{3} \times 500 \times 10.2 = 8.83\,kVAr$

EXAMPLE I.6
A 60 Hz, wye-connected synchronous generator has a 0.5 Ω/phase synchronous reactance and a negligible armature resistance. It is connected to a three-
phase 690 V (V_L) system and is exporting 300 kW at a power factor of 0.85
lagging. Calculate the line current, the internal voltage and the power angle
of the generator.

Answer:

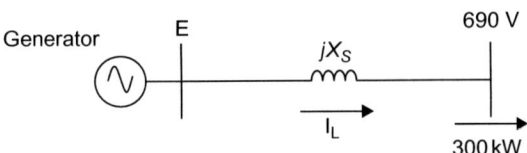

$P = \sqrt{3}V_L I_L\cos\phi$
$300 \times 10^3 = \sqrt{3} \times 690 \times I_L \times 0.85$

Therefore, the magnitude of the line current $= I_L = 295.32\,A$
The angle of the line current $= \cos^{-1}0.85 = 31.79°$
As the power factor is lagging $\mathbf{I_L} = 295.32\angle-31.8°$
From the figure, assuming that 690 V bus voltage is the reference phasor:

$\mathbf{E} = \mathbf{I_L} \times jX_s + \dfrac{690}{\sqrt{3}}\angle0°$

$= 295.32\angle-32.8° \times j0.5 + 398.37$

$= 147.66\angle57.2° + 398.37$

$= 80 + j124.12 + 398.37$

$= 494.2\angle14.55°\,V$

Internal voltage $= 494.2\,V$
Power angle $= 14.55°$

I.13   Problems

1. If $v = 10\sin(\omega t + \pi/3)$ and $i = 2\sin(\omega t - \pi/6)$, find (a) the rms value of voltage and current, (b) the phasor expression in polar form and (c) phase shift between the current and voltage.

(*Answers*: (a) 7.07 V, 1.41 A; (b) 7.07∠60°, 1.41∠ − 30° and (c) voltage leads the current by 90°)

2. A 60 μF loss-free capacitor is connected in parallel with a coil having an inductance of 0.1 H and a resistance of 10 Ω. The network is connected across an AC voltage $V = 10∠0°$ V at a frequency of 100 Hz. Draw the circuit and calculate the value of the current in each branch and its phase relative to the supply voltage. Sketch the calculated currents on a phasor diagram.

(*Answers*: Current through: capacitor 0.377 ∠90°, coil 0.157 ∠ − 81°)

3. A load having a series connected resistor of 5 Ω and a reactor of 2 Ω is connected across a 240 V AC supply.
   (a) Find the active and reactive power consumption of the load and its power factor.
   (b) What is the value of power factor correction capacitor in Farad required to improve the power factor to 0.95?

(*Answers*: (a) P = 9.93 kW, Q = 3.97 kVAr and power factor = 0.93; (b) 0.26 F)

4. A star-connected load having 10 Ω resistance per phase is connected to a three-phase 400 V supply. Calculate the current and power drawn by the load.

(*Answers*: I = 23.09 A; P − 5.33 kW)

5. A balanced delta-connected load of pure resistance of 10 Ω per phase is connected in parallel with a balanced star-connected load having impedance 5 + j2 Ω per phase. Both loads are connected to a 400 V three-phase supply. Calculate the current drawn by each load and hence calculate the total current drawn by the loads.

(*Answers*: current drawn by: delta load = 69.28∠0°, star load = 42.88∠ − 21.8°, total load = 110.26∠ − 8.3°)

I.14   Further reading

1. Hughes E., Hiley J., Brown K., McKenzie-Smith I. *Electrical and Electronic Technology*. Prentice Hall; 2008.
2. Grainger J.J., Stevenson W.D. *Power System Analysis*. McGraw-Hill; 1994.
3. Walls R., Johnston W. *Introduction to Circuit Analysis*. West Publishing Company; 1992.

AC machines

North Hoyle Offshore Wind Farm (60 MW)
This was the UK's first major offshore wind farm and is located 4–5 miles off the
North Wales coast. Each turbine is 2 MW [RWE npower renewables]

II.1 Introduction

AC electrical machines can act as generators to convert mechanical energy into electrical energy or motors to convert electrical energy into mechanical energy. For example, in a hydro power station, the kinetic and potential energy in the water are converted into mechanical rotational energy by the turbine and then into electricity by a generator. On the other hand, an AC motor in a factory provides mechanical energy when supplied with AC. In principle, an electrical machine of the same construction can be used as either a motor or a generator.

There are two major types of AC machine, synchronous and asynchronous (sometimes known as induction machines). In a synchronous machine, the rotor is supplied by DC or has permanent magnets mounted on it. The stator carries a three-phase winding. In an induction machine, both rotor and stator carry three-phase windings.

II.2   Synchronous machines

II.2.1   *Construction and operation*

1. Rotor construction and its magnetic field

Three forms or rotor constructions are commonly used in synchronous machines: permanent magnet, cylindrical rotor or salient pole, as shown in Figure II.1. The rotors of the cylindrical rotor and salient pole synchronous machine carry a field winding, which is fed from a source of DC. Even though a two-pole rotor structure is shown in this figure, two, four or even more poles are often used.

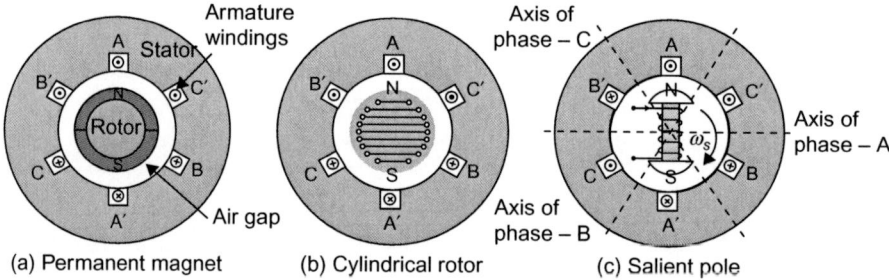

(a) Permanent magnet   (b) Cylindrical rotor   (c) Salient pole

Figure II.1   Rotor construction of synchronous machines

In cylindrical rotor and salient pole machines, the rotor is fed with DC through slip rings, as shown in Figure II.2.

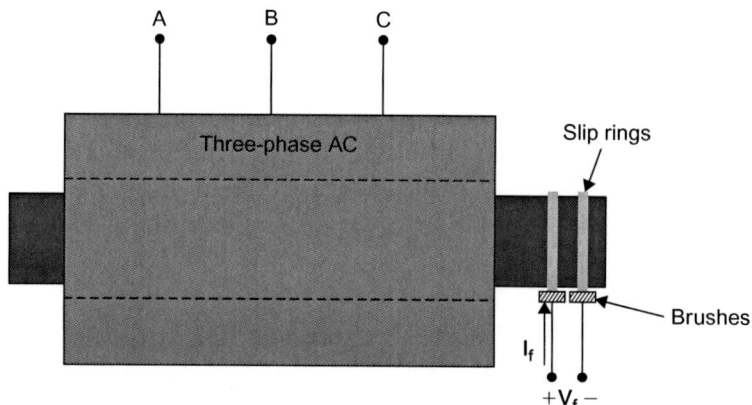

Figure II.2   Schematic representation of a synchronous machine

The DC on the rotor circuit (or the permanent magnets) produces a magnetic field, ϕ_{Rotor}, that is fixed to the rotor. As the rotor rotates at synchronous speed (ω_s), the rotor magnetic field also rotates at the same speed.

2. Stator construction and its magnetic field
    Figure II.1 also shows the stator of an AC machine. The stator core consists of slotted ring shape laminations that are stacked and bolted together to form a cylindrical core. The slots on this cylindrical structure carry the armature winding. Even though only three armature windings (A–A', B–B' and C–C') are shown in Figure II.1 for simplicity, in a real machine the stator has several armature coils in each phase.
    When the rotor rotates, the magnetic field produced by the field winding sweeps the three-phase armature windings. This in turn induces three voltages in three windings, A–A', B–B' and C–C' (initially assumed to be open circuit), which are displaced by 120°. At the position shown in Figure II.1(c), the induced voltage of the A phase is at its positive maximum and as the rotor rotates its magnitude reduces. Once the rotor rotates by 180° the induced voltage comes to its negative maximum. Hence, the frequency of the induced voltage is directly proportional to the speed of rotation of the rotor and is given by $f = \omega_s/2\pi$.

If the armature windings are connected to a balanced three-phase load, the resulting three-phase currents are also displaced by 120° as shown in Figure II.3.

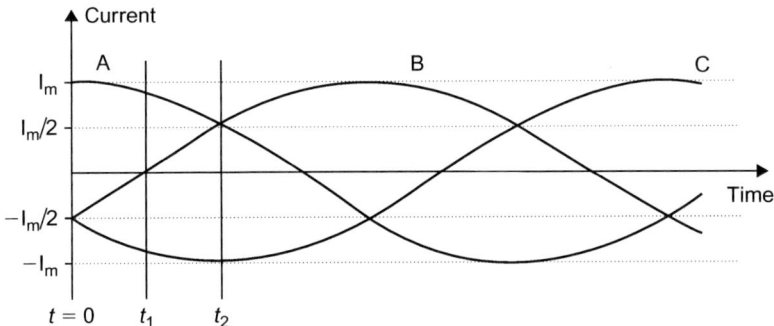

Figure II.3   Three-phase currents ($t_1 = \pi/6\omega$ and $t_2 = \pi/3\omega$)

When $t = 0$ (Figure II.3) the current in phase A is I_m and the currents in phases B and C are equal to $-(I_m/2)$. The currents in phases A, B and C produce components of the stator magnetic field, ϕ_A, ϕ_B and ϕ_C. The magnitudes of these flux components are proportional to the number of ampere-turns, NI_m, $NI_m/2$ and $NI_m/2$ (where N is the effective number of turns of each phase of the stator winding) and act along the axis of phases A, B and C (see Figures II.1(c) and II.4). These three magnetic fields combine and produce a resultant stator magnetic field, ϕ_{Stator}, at $t = 0$ as shown in Figure II.4. The distribution of this flux in the air gap is also shown in

Figure II.5(a). Similarly, the resultant magnetic fields produced by the stator current at $t = t_1$ and t_2 are shown in Figure II.5(b) and (c).

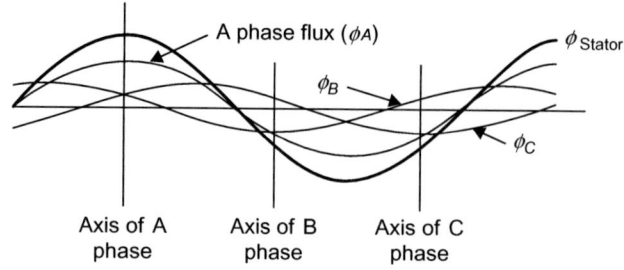

Axis of A   Axis of B   Axis of C
phase   phase   phase

Figure II.4   Resultant stator magnetic field in the air gap at t = 0 (The x axis shows the air gap, around the circumference of the machine.)

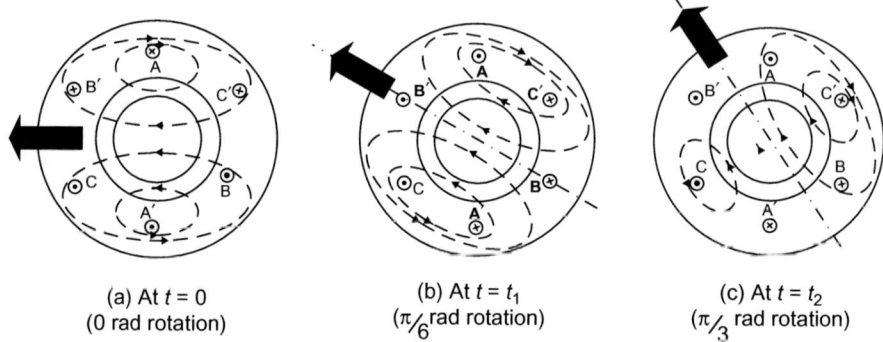

(a) At $t = 0$ (b) At $t = t_1$ (c) At $t = t_2$
(0 rad rotation)   ($\pi/6$ rad rotation)   ($\pi/3$ rad rotation)

Figure II.5   Illustration of the rotating stator magnetic field

During the time $t_1 = (\pi/6\omega_s)$, the flux waveform has moved by $\pi/6$. This is the speed of the stator magnetic field and is equal to synchronous speed. In normal operation the rotor, and hence the field winding, rotates synchronously with the flux developed by the stator. The relative angle between the rotor flux axis and the stator field, the rotor angle (sometimes known as the load angle), is determined by the torque applied to the shaft.

II.2.2   Electrical and mechanical angle

In the case of a two-pole synchronous machine, for 180° of rotation of the rotor the induced voltage changes by half a cycle or 180° electrical (Π radians). Now consider the four-pole machine shown in Figure II.6. At the rotor position shown, the A phase induced voltage is at positive maximum. Once the rotor rotates by 90°, a south pole comes immediately below the A phase and the induced voltage becomes negative maximum. Hence, a 90° mechanical rotation corresponds to 180° electrical variation in the induced voltage.

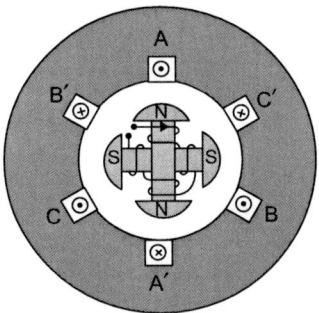

Figure II.6   Four-pole synchronous machine

In general if there are p number of poles, the relationship between the electrical (θ_e) and mechanical (θ_m) angles is given by:

$$\theta_e = \frac{p}{2}\theta_m \qquad\qquad\qquad (\text{II.1})$$

By dividing both sides of (II.1) by time, one can obtain a relationship between the mechanical and electrical rotational speeds:

$$\omega_e = \frac{p}{2}\omega_m \qquad\qquad\qquad (\text{II.2})$$

Assuming that the torque applied on the rotor by a turbine is T_m and electromagnetic torque is T_e, then from power balance:

$$T_m\omega_m = T_e\omega_e \qquad\qquad\qquad (\text{II.3})$$

From (II.2) and (II.3):

$$T_m = \frac{p}{2}T_e \qquad\qquad\qquad (\text{II.4})$$

Direct-drive wind turbine generators are directly connected to the aerodynamic rotor. Hence, they have large number of poles and operate a low mechanical rotational speed but with high mechanical torque.

II.2.3   Equivalent circuit

To investigate how a synchronous generator will behave on the power system, a simple model is required. When the stator is open circuit, the stator terminal voltage is equal to the internal voltage, $\mathbf{E_F}$, as shown in Figure II.7.

When the armature windings are connected to a three-phase load, the magnetic field produced by the armature currents interacts with the rotor magnetic field, ϕ_{Rotor}. The effect of ϕ_{Stator} on ϕ_{Rotor} is called armature reaction. The armature reaction is normally represented by a reactance in the synchronous machine equivalent circuit.

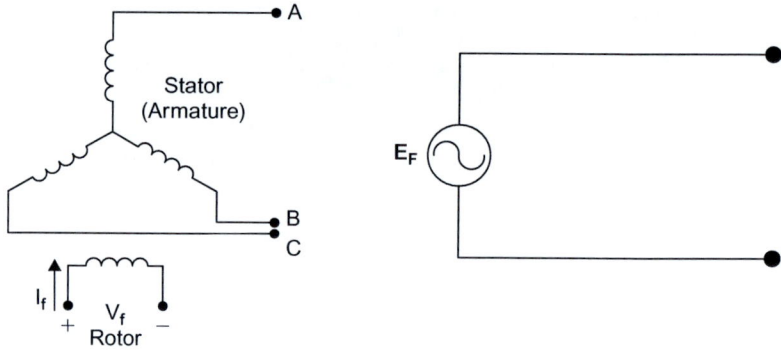

Figure II.7   Synchronous machine on open circuit

Further, a part of the flux produced by the rotor is not linked with the stator and that component is called leakage flux. This component is also represented by a reactance. Finally voltage drop across the stator resistance is also taken into account when developing the equivalent circuit. The equivalent circuit is shown in Figure II.8.

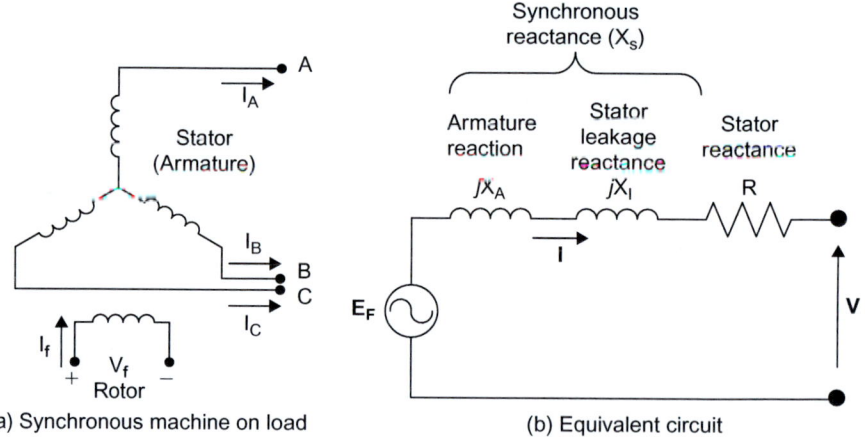

(a) Synchronous machine on load   (b) Equivalent circuit

Figure II.8   Synchronous machine equivalent circuit

Generally for synchronous generators, X_s is much greater than R and therefore R is neglected. From Figure II.8(b), it may be seen that:

$$\mathbf{V} = \mathbf{E_F} - j\mathbf{I}X_s \tag{II.5}$$

where $\mathbf{V}$ = terminal voltage, $\mathbf{E_F}$ = internal voltage (function of the field current), X_s = synchronous reactance.

For a small distributed generator, the terminal voltage is held almost constant by the network and so phasor diagrams may be developed (Figure II.9) that illustrate the operation of a synchronous generator on to a fixed voltage (or infinite busbar).

The power factor of the power delivered to the network is simply cos ϕ, while the rotor or load angle (the angle by which the rotor is in advance of the stator voltage) is given by δ.

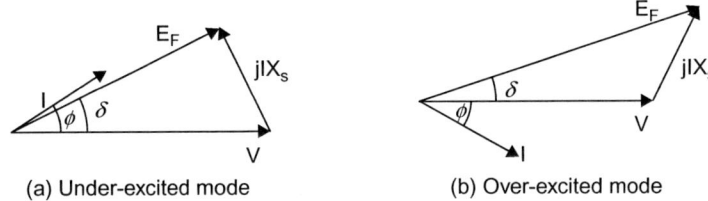

(a) Under-excited mode   (b) Over-excited mode

Figure II.9   Phasor diagram of synchronous generator (ϕ: power factor angle, δ: rotor angle)

From Figure II.9(b):

$$I = \frac{E_F - V}{jX_s} = \frac{E_F\angle\delta - V\angle 0}{jX_s} = \frac{E_F \sin\delta}{X_s} - j\frac{E_F \cos\delta - V}{X_s}$$

$$S = VI^*$$

$$= V \times \frac{E_F \sin\delta}{X_s} + jV \times \frac{E_F \cos\delta - V}{X_s}$$

$$= P + jQ$$

Therefore:

$$P = \frac{E_F V \sin\delta}{X_s} \tag{II.6}$$

$$Q = \frac{E_F V \cos\delta - V^2}{X_s} \tag{II.7}$$

In normal operation, the rotor angle δ is usually less than 30°. Hence the real power output (P) is proportional to the rotor angle (δ). Increasing the torque on the rotor shaft increases the rotor angle (δ) and results in more active power exported to the network, as shown in (II.6). As the rotor angle is a function of the load on the rotor shaft, it is also known as the load angle.

Again with a rotor angle of less than 30°, cos (δ) remains approximately constant. Increasing the field current and hence increasing the magnitude of $\mathbf{E_F}$ results in export of reactive power, as shown in (II.7).

The phasor diagrams of Figure II.9 show two different values of excitation (determined by the magnitude of E_F).

1. Under-excited $E_F < V$
   This gives a leading power factor (using a generator convention and the direction of **I** as shown in Figure II.8(b)), importing reactive power.

2. Over-excited: $E_F > V$
    This gives a lagging power factor (using a generator convention and the direction of **I** as shown in Figure II.8(b)), exporting reactive power.

It may be noted that if the direction of the definition of the current **I** is reversed, and the machine considered as a motor rather than a generator, then an under-excited motor has a lagging power factor and an over-excited motor has a leading power factor. Of course, if torque is still applied to the shaft, then active power will be exported to the network and if $E_F > V$ then reactive power will still be exported irrespective of whether the same machine is called a motor or a generator. Therefore, it is often helpful to consider export/import of real and reactive power than leading/lagging power factors that rely on the definition of the direction of the current flow.

II.2.4   *Operating chart of a synchronous generator*

The operating chart of a synchronous generator is formed directly from the phasor diagram of Figure II.9. The phasor diagram is simply scaled by multiplying by V/X_s, which is a constant, to give the phasor diagram of Figure II.10. The locus of the new phasor **VI** then describes the operation of the generator. Various limits are applied to account: (1) for the maximum power available from the prime mover, (2) the maximum current rating of the stator, (3) the maximum excitation and (4) the minimum excitation for stability and/or stator end winding heating. These limits then form the boundaries of the region within which a synchronous generator may operate. In practice, there may be additional limits including a minimum power requirement of the prime mover and the effect of the reactance of the generator transformer.

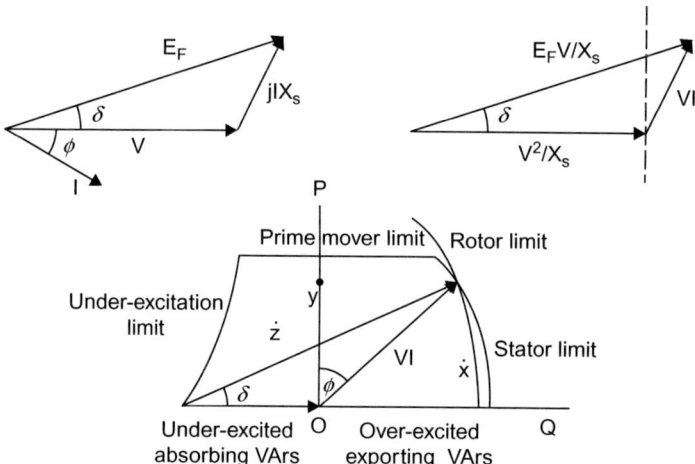

Figure II.10   Operating chart of a cylindrical pole synchronous generator connected to an infinite bus

The operating chart illustrates that a synchronous generator connected to an infinite busbar of fixed voltage and frequency has essentially independent control over real and reactive power. Real power is varied by adjusting the torque on the generator shaft and hence the rotor angle, while reactive power is adjusted by varying the field current and hence the magnitude of $\mathbf{E_F}$. For example at point (x) both real and reactive power are exported to the network, at (y) rather more real power is being exported at unity power factor, while at (z) real power is exported and reactive power is imported.

II.2.5 Excitation systems

The performance of a synchronous generator is strongly influenced by its excitation system particularly its transient and dynamic stability, and the ability of the generator to deliver sustained fault current.

On some older generators, a DC generator, with a commutator, was used to provide the field current, which was then fed to the main field via slip rings on the rotor. Equipment of this type can still be found in service, although often with modern AVRs replacing the rather simple voltage regulators that were used to control the field of the DC generator and hence the main excitation current. However, more modern excitation systems are generally of two types: (1) brushless or (2) static.

Figure II.11 is a schematic representation of a brushless excitation system. The exciter is simply an alternator, much smaller than the main generator and with a stationary field and a rotating armature. A full wave diode bridge is mounted on the rotating shaft to rectify the three-phase output of the exciter rotor to DC for the field of the main generator. The exciter field is controlled by the AVR that is set to control either the generator terminal voltage or, using a power factor controller, to maintain either a constant power factor or a defined reactive power output. Power for the exciter may be taken either from the terminals of the main generator (self-excited) or from a permanent magnet generator (separately excited). The permanent magnet generator is mounted on an extension of the main generator shaft and continues to supply power as long as the generator is rotating.

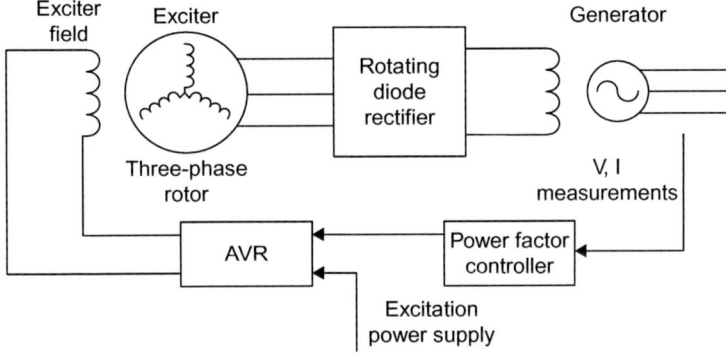

Figure II.11   Brushless excitation system

Figure II.12 shows a static excitation system in which DC is supplied by a controlled thyristor rectifier and fed to the generator field via slip rings. The power supply for the thyristor rectifier is taken from the terminals of the generator. The main advantage of the static exciter is improved speed of response as the field current is controlled directly by the thyristor rectifier but, of course, if the generator terminal voltage is depressed too low then excitation power will be lost.

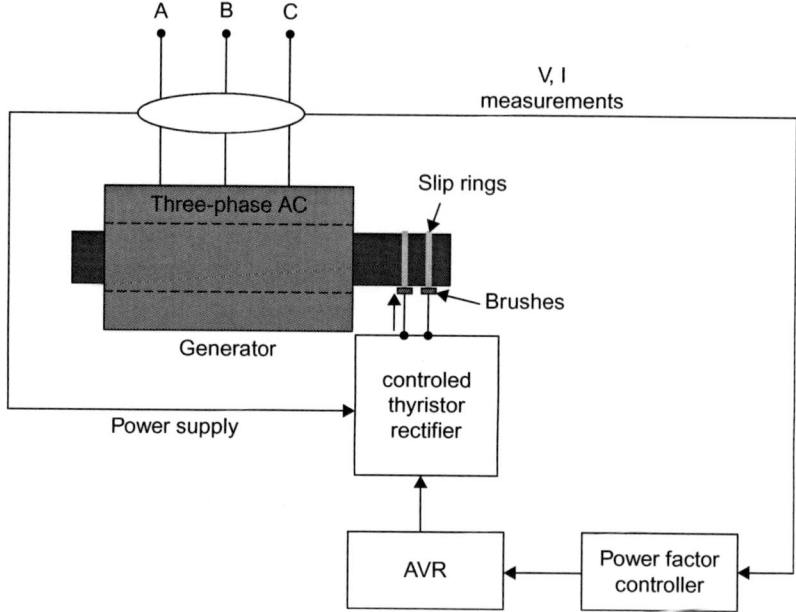

Figure II.12   Static excitation system

In addition to the main types described, there are a large number of innovative designs of excitation systems, which have been developed over the years particularly for smaller generators. These include the use of magnetic circuits for the no-load excitation and current transformer compounding for the additional excitation required as current is drawn. Although such techniques may work robustly on standalone systems, they are almost impossible to model for studies of distributed generation schemes. For larger generators and their excitation systems the manufacturers are usually able to supply the so-called IEEE exciter models. These refer to the structure of excitation system models that have been developed by the IEEE and are included in most power system analysis programs.

EXAMPLE II.1

A 13.8 kV(V_L), 20 MVA, 50 Hz, three-phase, star-connected synchronous generator has a synchronous reactance of 1.5 Ω/phase and negligible resistance. It supplies 10 MW at a power factor of 0.8 lagging (exporting VArs) to an infinite busbar held at 13.8 kV. What is the internal voltage and power angle of the generator?

Answer:
The line voltage of the infinite busbar is 13.8 kV and $P = \sqrt{3}V_L I_L \cos \phi$

Hence $I_L = \dfrac{P}{\sqrt{3}V_L \cos \phi} = \dfrac{10 \times 10^6}{\sqrt{3} \times 13.8 \times 10^3 \times 0.8} = 523$ A

Angle of load current $= \cos^{-1}(0.8) = 36.87°$

Since the power factor is lagging and the generator exporting VArs:
$\mathbf{I_L} = 523\angle - 36.87°$

From (II.5):

$$\mathbf{E_F} = \mathbf{V} + j\mathbf{I}X_s$$

$$= \frac{13.8 \times 10^3}{\sqrt{3}} \angle 0° + 523\angle - 36.87° \times j1.5$$

$$= 7967.4 + 784.5\angle 53.13°$$

$$= 7967.4 + 470.7 + j627.6$$

$$= 8438.1 + j627.6$$

$$= 8461.4\angle 4.25°$$

Therefore, the magnitude of the generator internal voltage $= 8461.4 \times \sqrt{3} = 14.66\,\text{kV}$ Rotor angle $= 4.25°$.

EXAMPLE II.2

A 12 kV(V_L), 120 MVA, 50 Hz, three-phase, star-connected synchronous generator has a synchronous reactance of 0.3 Ω/phase and negligible resistance. It is connected to an infinite busbar and supplies 60 MW at power factor of 0.85 lagging. If the excitation of the machine is increased by 20% and the mechanical power input by 25%, determine the subsequent rotor angle at which the machine operates.

Answer:

The line voltage of the infinite busbar is 12 kV, and the generator supplies 60 MW at a lagging power factor of 0.8.

$$I_L = \frac{P}{\sqrt{3}V_L \cos\phi} = \frac{60 \times 10^6}{\sqrt{3} \times 12 \times 10^3 \times 0.85} = 3396.2\,\text{A}$$

Angle of load current $= \cos^{-1}(0.85) = 31.8°$
Since the load is lagging: $\mathbf{I_L} = 3396.2\angle -31.8°$
From (II.5):

$$\mathbf{E_F} = \mathbf{V} + j\mathbf{I}X_s$$

$$= \frac{12 \times 10^3}{\sqrt{3}} \angle 0° + 3396.2\angle - 31.8° \times j0.3$$

$$= 6928.2 + 1018.86\angle 58.2°$$

$$= 6928.2 + 536.9 + j865.9$$

$$= 7515.15\angle 6.6°$$

If the efficiency of the generator is η, the input mechanical power $= 60/\eta$ MW

If the mechanical power is increased by 25%, the new mechanical power input $= 1.25 \times 60/\eta$ MW

New electrical output $= 1.25 \times (60/\eta) \times \eta = 75$ MW

When the excitation increases the internal voltage by 20% and the mechanical input is increased by 25%, from (II.6):

$$75 \times 10^6 = \frac{1.2 \times 7515.15 \times 6928.2 \times \sin \delta_2}{0.3}$$

$$\sin \delta_2 = 0.36$$

$$\delta_2 = 21.1°$$

New rotor angle is 21.1°.

II.3   Induction machines

II.3.1   *Construction and operation*

The stator of an induction machine is a laminated structure (similar in construction to a synchronous machine stator), which carries a three-phase winding and is fed with a three-phase power supply.

The commonly used rotor is a squirrel-cage construction as shown in Figure II.13(a). In this rotor, solid copper or aluminium bars are embedded in a laminated rotor structure and short circuited by two end rings (Figure II.13(a) – bottom). The other rotor construction is called wound rotor, where the rotor carries three-phase

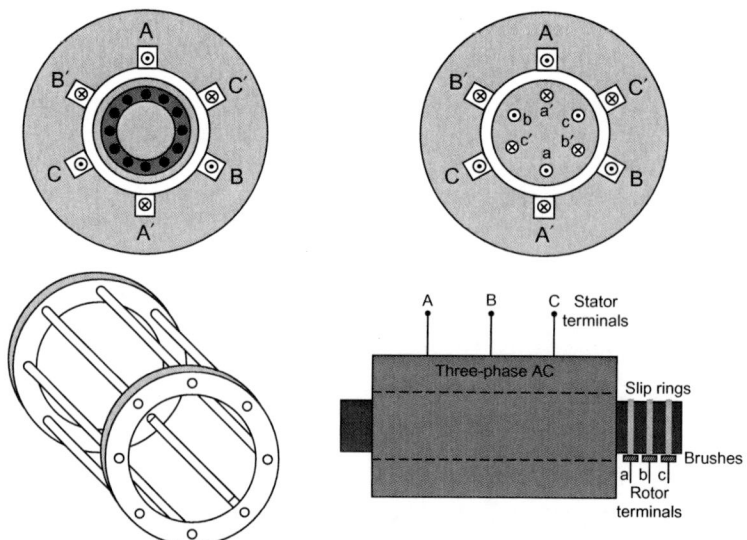

(a) Squirrel-cage induction machine   (b) Wound-rotor induction machine

Figure II.13   Induction machine construction

windings as shown in Figure II.13(b). The terminals of the three windings, connected in star or delta, are taken out through slip rings ((Figure II.13(b) – bottom). In wound-rotor induction motors, and some variable speed wind turbines, these windings are often short circuited through a set of resistors.

To describe the operation of the induction machine, it is easier to start with motoring operation. When the stator windings are connected to a three-phase supply, a rotating magnetic field is set up as described for a synchronous machine. This rotating magnetic flux (ϕ_{Stator}) cuts the rotor conductors, which are stationary at start-up, and induces a voltage. Since the rotor consists of three-phase windings (or a squirrel cage that forms a three-phase winding), the induced voltage in each rotor phase will be displaced in space by 120°. Normally in induction machines, all three phases in the rotor are short circuited and therefore the induced voltage in the rotor produces a circulating current. Three-phase currents flowing in the rotor will also produce a rotating magnetic field (ϕ_{Rotor}). There will be an alignment force between stator and rotor magnetic fields, thus creating a torque proportional to:

$$T \propto \phi_{Stator} \times \phi_{Rotor} \times \sin \theta \qquad (\text{II.8})$$

where θ is the angle between the two fluxes.

The rotor then accelerates to its running speed (ω_r) slightly less than synchronous speed, ω_s. Since $\omega_r < \omega_s$, there is still relative movement between the rotor conductors and the stator flux, thus maintaining the rotor current and flux. The normalised value of the difference between the running speed of the rotor and the synchronous speed is defined as the slip and given by:

$$s = \frac{\omega_s - \omega_r}{\omega_s} \qquad (\text{II.9})$$

II.3.2 Steady-state operation

An induction machine can be considered as a transformer where the stator acts as the primary and the rotor acts as the secondary. The main difference between a transformer and an induction machine is that the frequency of the induced voltage on the rotor circuit will differ from the stator frequency as the rotor rotates.

When the rotor is not rotating ($\omega_r = 0$):
The relative motion between the rotor conductors and the stator magnetic field is ω_s. Assume that the rotor-induced voltage is equal to $\mathbf{E_2}$ and rotor inductance is L_2. As the rotor-induced voltage is proportional to the speed of rotation of the rotor conductors with respect to the stator magnetic field:[1]

$$E_2 \propto \omega_s \qquad (\text{II.10})$$

The frequency of $\mathbf{E_2}$ is $\omega_s/2\pi = f$ $\qquad (\text{II.11})$

Rotor reactance $X_2 = 2\pi f L_2$ $\qquad (\text{II.12})$

---

[1] From Faraday's law: $E = -N(d\phi/dt)$. If ϕ is sinusoidal as shown in Figure II.4, then $\phi = \phi_m \sin \omega_s t$. Therefore, $E = -N\phi_m \omega_s \sin \omega_s t$; $E \propto \omega_s$.

When the rotor is rotating at ω_r:

There will be a relative motion between the rotor conductors and the stator magnetic field equal to $\omega_s - \omega_r = s\omega_s$ (see (II.9)). If the rotor-induced voltage is $\mathbf{E}_2^r$:

$$E_2^r \propto s\omega_s \tag{II.13}$$

The frequency of $\mathbf{E}_2^r$ is $s\omega_s/2\pi$ \hfill (II.14)

From (II.10) and (II.13), $E_2^r = sE_2$ and from (II.11) and (II.14), the frequency of the rotor-induced voltage when the rotor is running is sf
If the rotor reactance is X_2^r:

$$X_2^r = 2\pi sfL_2 \tag{II.15}$$

From (II.12) and (II.15), $X_2^r = sX_2$.

Figure II.14 shows the transformer equivalent circuit of an induction machine when the rotor is running with slip s.

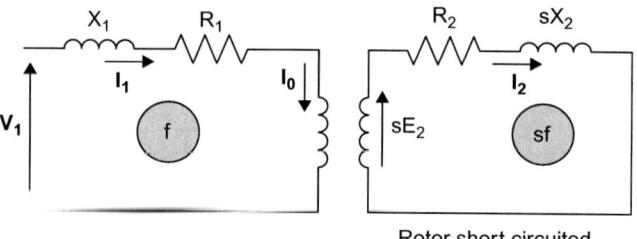

Rotor short circuited

Figure II.14   Induction machine equivalent circuit

In Figure II.14, R_1 is the stator resistance, X_1 is the stator leakage reactance, R_2 is the rotor resistance and X_2 is the rotor leakage reactance at standstill.

From the rotor side of the equivalent circuit:

$$I_2 = \frac{sE_2}{R_2 + jsX_2} \tag{II.16}$$

Dividing both the numerator and denominator by s:

$$I_2 = \frac{E_2}{(R_2/s) + jX_2} = \frac{E_2}{R_2 + R_2((1/s) - 1) + jX_2} \tag{II.17}$$

Now by transforming the rotor quantities into the stator and by considering (II.17), the stator referred equivalent circuit of an induction machine is obtained. This is the familiar (Steinmetz) induction motor equivalent circuit and is shown in Figure II.15.

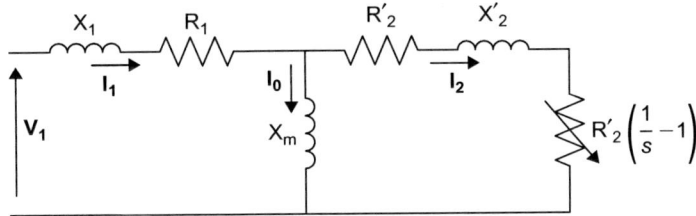

Figure II.15 Equivalent circuit of the induction machine

In Figure II.15, R'_2 is the rotor resistance referred to the stator side, X'_2 is the rotor leakage reactance referred to the stator side, X_m is the magnetising reactance (represents the current required to set up the air-gap flux) and $R'_2((1-s)/s)$ represents the mechanical load in the case of a motor or turbine input in the case of a generator.

The usual simple analysis of this circuit relies either on moving the magnetising branch to the supply terminals (the so-called approximate equivalent circuit) or by using a Thevenin transform to eliminate the shunt branch.

Considering the approximate equivalent circuit, the current flowing in the rotor circuit is given simply by:

$$I_2 = \frac{V}{(R_1 + (R'_2/s)) + j(X_1 + X'_2)} \tag{II.18}$$

The total power supplied to the rotor (from the stator through the air gap) is the sum of the copper losses and the developed mechanical power.

If the electromagnetic torque is T_e, the power transferred to the rotor is:

$$P_{\text{air gap}} = T_e \omega_s \tag{II.19}$$

The mechanical power of the rotor is given by:

$$P_{\text{mech}} = T_e \omega_r \tag{II.20}$$

The rotor copper losses are given by:

$$P_{\text{losses}} = 3I_2^2 R'_2 \tag{II.21}$$

From power balance considerations:

$$T_e \omega_s = T_e \omega_r + 3I_2^2 R'_2$$
$$T_e(\omega_s - \omega_r) = 3I_2^2 R'_2 \tag{II.22}$$

$$\therefore T_e = \frac{3I_2^2 R'_2}{s\omega_s} \tag{II.23}$$

Substituting for I_2 from (II.18) into (II.23):

$$T_e = \frac{3V^2 R_2'}{s\omega_s[(R_1 + (R_2'/s))^2 + (X_1 + X_2')^2]}$$ (II.24)

Equation (II.24) then allows the familiar torque-slip curve of an induction machine to be drawn (Figure II.16). Although the steady-state equivalent circuit is extremely useful in understanding the performance of induction generators, it should be remembered that it is an approximate model only. The simple steady-state equivalent circuit model ignores any effects due to harmonics and magnetic saturation. It is not suitable for investigating transient behaviour and does not take into account rotor frequency effects.

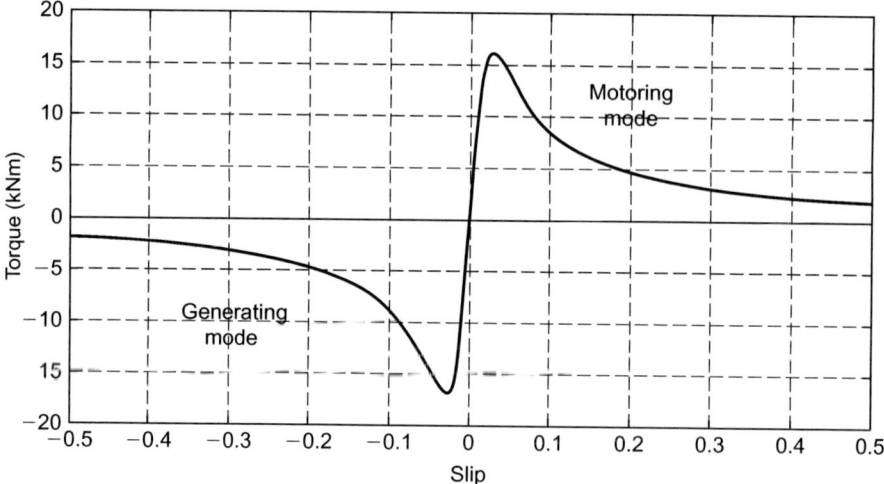

Figure II.16   Torque–slip curve for a directly connected 2 MW induction machine

Figure II.16 shows that for this example (2 MW, 690 V induction generator) the pull-out torques in both the motoring and generating regions are in excess of 200% (100% torque $= 2 \times 10^6/314.16 = 6.37$ kNm, where 314.16 is the speed in rad/s for a two-pole induction generator). The normal operating locus of an induction machine may also be described in terms of real and reactive power (in a similar manner to the synchronous machine operating chart of Figure II.10). This is the well-known circle diagram and is shown in Figure II.17. Compared to a synchronous machine, the major difference is that an induction generator can only operate on the circular locus and so there is always a defined relationship between real and reactive power. Hence, independent control of the power factor of the output of a simple induction generator is not possible. For example, at point (B) the generator is exporting active power but importing reactive power, while at point (A) no power is exported but the no-load reactive power is absorbed. It may be seen that the power factor decreases with reducing load.

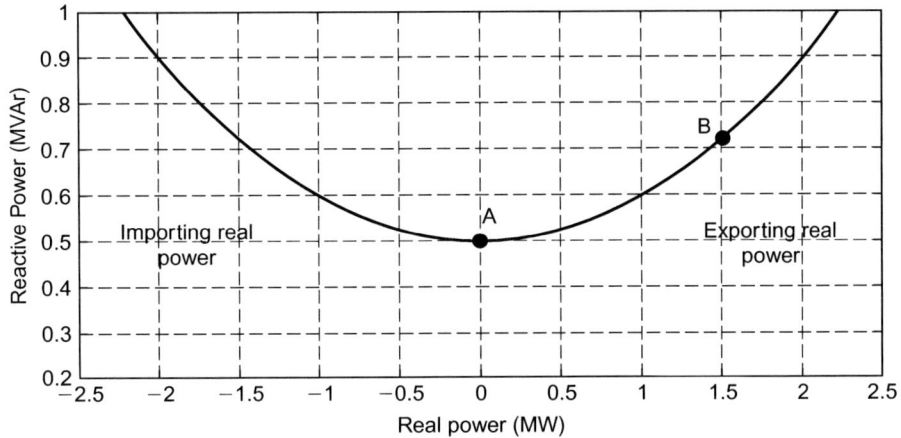

Figure II.17 Circle diagram of the 2 MW induction generator

EXAMPLE II.3
The speed of a 400 V, 35 kW, 50 Hz, four-pole induction motor at the rated load is 1455 rev/min. What is the

(a) slip of the rotor?
(b) frequency of the rotor current?

Answer:
(a) The speed of the stator magnetic field = 2 × π × 50 = 314.16 rad/s
     From (II.2), the corresponding mechanical synchronous speed is: ω_s = $(2/p)\omega_e$ = (2/4) × 314.16 = 157.08 rad/s
     The rotor speed, ω_r = 1455 × (2π/60) = 152.37 rad/s
     Therefore, the slip of the inductor motor = (157.08–152.37)/157.08 = 0.03 = 3%
(b) The frequency of the rotor current = sf = 0.03 × 50 = 1.5 Hz

---

EXAMPLE II.4
A three-phase 400V(V_L), 50 Hz induction machine has a stator impedance of 0.05 + j0.41 Ω/phase, a rotor impedance referred to the stator side of 0.055 + j0.42 Ω/phase and a magnetising reactance of j8 Ω/phase.

(a) Find the slip at the maximum torque and hence the maximum slip in both motoring and generating regions. Sketch the shape of the torque-slip curve indicating the motoring and generating regions. Mark the normal operating regions of an induction generator.

(b) Calculate the total no-load reactive power that is drawn (at s = 0). (The magnetising reactance may be moved to the terminals of the circuit.)

(c) Calculate the size (in F) of the power factor correction capacitors that would be connected across the motor terminals to give a no-load power factor of 1.

(d) The machine operates at a slip of −0.04. Calculate the current in the rotor circuit and hence the real and reactive power flows in the network connection (assuming the capacitor calculated in (c) is in place.)

Answer:

$R_1 = 0.05\,\Omega$, $X_1 = 0.41\,\Omega$, $R_2' = 0.055\,\Omega$, $X_2' = 0.42\,\Omega$, $X_m = 8\,\Omega$ and phase voltage $= (400/\sqrt{3}) = 230.9\,V$.

(a) From (II.24):

$$T_e = \frac{3V^2R_2'}{s\omega_s[(R_1 + (R_2'/s))^2 + (X_1 + X_2')^2]} = \frac{3V^2R_2'}{\omega_s \times g(s)}$$

For torque to be maximum, the function $g(s)$ should be minimum. That is $d[g(s)]/ds = 0$

$$\frac{d[g(s)]}{ds} = \left[\left(R_1 + \frac{R_2'}{s}\right)^2 + (X_1 + X_2')^2\right] + s\left[2 \times \left(R_1 + \frac{R_2'}{s}\right) \times \frac{R_2'}{s^2}\right] = 0$$

$$\left[R_1^2 - \left(\frac{R_2'}{s}\right)^2 + (X_1 + X_2')^2\right] = 0$$

$$s = \pm \frac{R_2'}{\sqrt{R_1^2 + (X_1 + X_2')^2}}$$

Therefore, slip at maximum torque $=$ $s = \pm 0.055/$
$\sqrt{0.05^2 + (0.41 + 0.42)^2} = \pm 0.066$

At $s = 0.066$,

$$T_e = \frac{3 \times 230.9^2 \times 0.055}{314.16 \times 0.066 \times [(0.05 + (0.055/0.066))^2 + (0.41 + 0.42)^2]}$$
$$= 288.8\,Nm$$

At $s = -0.066$,

$$T_e = -\frac{3 \times 230.9^2 \times 0.055}{314.16 \times 0.066 \times [(0.05 - (0.055/0.066))^2 + (0.41 + 0.42)^2]}$$
$$= -325.7\,Nm$$

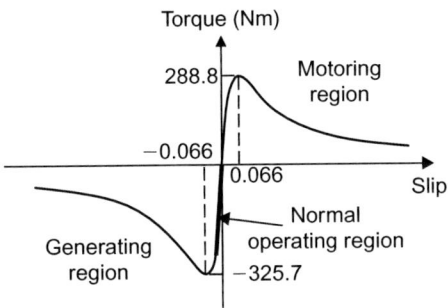

(b) From Figure II.15, on no-load (i.e. with s = 0) one can see that the rotor circuit is open circuited. When the magnetising reactance is moved to the terminals:

The reactive power drawn by one phase = $V^2/X_m = 230.9^2/8 = 6.67$ kVAr

Therefore, no-load reactive power = $3 \times 6.67 = 20$ kVAr

(c) If the per phase capacitor connected to provide no-load reactive power completely (thus unity power factor operation) is C:

$$V^2\omega C = 2 \times \pi \times 50 \times C \times 230.9^2 = 6.67\,\text{kVAr}$$

$$\therefore C = \frac{6.67 \times 10^3}{2 \times \pi \times 50 \times 230.9^2} = 398\mu F$$

(d) At a slip of -0.04, from (II.18), the stator current is given by:

$$= \frac{230.9}{(0.05 - (0.055/0.04)) + j(0.41 + 0.42)}$$

$$= -125.15 - j78.4\,\text{A}$$

The apparent power in the stator circuit = $3VI^* = 3 \times 230.9 \times (-125.15 + j78.4) = -86.7 + j54.3$ kVA

The real power exported to the network = 86.7 kW.

The reactive power import is 54.3 kVAr in the stator circuit plus 20 kVAr in the magnetising branch minus 20 kVAr supplied by the capacitor. That is reactive power import is 54.3 kVAr.

II.4 Problems

1. A 11 kV, 50 Hz, three-phase, star-connected synchronous generator has an armature resistance of 0.05 Ω and a synchronous reactance of 1.2 Ω/phase. The

generator supplies 5 MW at 0.9 power factor lagging to an infinite busbar held at 11 kV. Determine the internal voltage and power angle required.

(*Answers*: Internal voltage = 11.3 kV and load angle = 2.7°)

2. A 11 kV, three-phase, star-connected synchronous generator has a synchronous reactance of 5 Ω/phase and negligible resistance. It delivers 1 MW at 0.9 power factor lagging. Find the rotor angle. If the load power is increased to 2 MW without changing the excitation, calculate the new value of rotor angle.

(*Answers*: Rotor angle with 1 MW load = 2.3°;
load angle with 2 MW load = 14°)

3. A synchronous generator has a synchronous reactance of 2 Ω/phase and negligible resistance. It is connected to a 6.6 kV infinite busbar and delivers 100 A at 0.9 power factor lagging. If the excitation is increased by 25%, what is the subsequent maximum power that the generator can deliver?

(Hint: Maximum power occurs when the rotor angle is 90°)

(*Answer*: 9.3 MW)

4. A 400 V, 50 Hz, four-pole, three-phase induction motor develops mechanical torque of 30 Nm, when running at 1440 rev/min. The mechanical losses are 250 W. Calculate the slip, electromagnetic torque and power transferred through the air gap.

(*Answers*: Slip = 4%, electromagnetic torque = 31.66 Nm,
air-gap power = 4.97 kW)

5. A three-phase induction motor has a stator leakage impedance of 1.0 + j4.0 Ω/phase and rotor leakage impedance of 1.2 + j4.0 Ω/phase. Determine the ratio of starting torque to maximum torque.

(*Answer*: 0.32)

II.5   Further reading

1. Hughes E., Hiley J., Brown K., McKenzie-Smith I. *Electrical and Electronic Technology*. Prentice Hall; 2008.
2. Chapmen S.J. *Electrical Machinery Fundamentals*. McGraw-Hill; 2005.
3. Hindmarsh J. *Electrical Machines and Their Applications*. Pergamon Press; 1994.

Tutorial III

Power electronics

Dun Law Wind Farm, Oxton, Scotland (47 MW)
Turbines are pitch regulated, variable speed using Doubly Fed Induction
Generators [RES]

III.1 Introduction

Power electronic converters are presently used to interface many forms of renewable generation and energy storage systems to distribution networks, while the use of power electronics is likely to increase in the future as this technology is also an important element of SmartGrids and active distribution networks. The development of high power electronic converters benefits from recent rapid advances in power semiconductor switching devices and in the progress being made in the design and control of variable speed drives for large motors.

One obvious application of a power electronic converter is to invert the DC generated from some energy sources (e.g. photovoltaics, fuel cells or batteries) to 50/60 Hz AC. Converters may also be used to de-couple a rotating generator and prime mover from the network and so allow it to operate at its most effective speed over a range of input powers. This is one of the arguments put forward in favour of the use of variable speed wind turbines but is also now being proposed for some small hydro generation. Another advantage of variable speed operation is the reduction in mechanical loads possible by making use of the flywheel effect to store energy during transient changes in input or output power.

However, large power electronic converters do have a number of disadvantages including: (1) significant capital cost and complexity, (2) electrical losses (which may include a considerable element independent of output power) and (3) the possibility of injecting harmonic currents into the network.

III.2   Conductors, insulators and semiconductors

Depending on their electrical conduction properties, elements are divided into three main categories: conductors, insulators and semiconductors.

III.2.1   Conductors

Elements through which electricity conducts easily are called conductors. Metals such as copper, silver and aluminium are good conductors. In these elements, the electrons in the outer orbits, which are called the valance electrons, are loosely bonded to the nucleus. Each valance electron inside a conductor has a different energy level, thus their cumulative energy level is represented by a band, the valance band (Figure III.1). When external energy in the form of heat, electricity or light is applied, these electrons break from the nucleus and move to the conduction band. These are now free electrons that can move easily when subjected to a small electric field. In a conductor the valance and conductance bands overlap.

A free electron migrates from one atom to another and replaces a valance electron in the second atom while leaving a positive charge on the first one. This movement of free electrons provides the electric current inside the conductor. Conventionally the direction of current flow is considered as being in the opposite direction to the electron flow and is in the direction of the movement of positive ions.

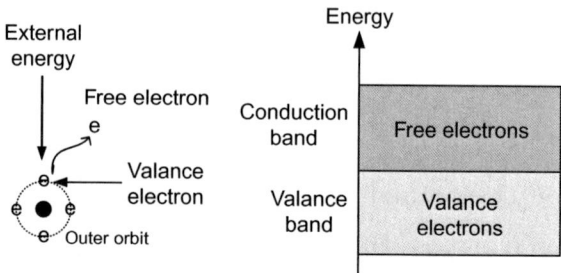

Figure III.1   Energy bands of a conductor

III.2.2 Insulators

The elements that do not conduct under normal conditions are called insulators. In good insulators the valance electrons are tightly bonded to the nucleus and therefore a large amount of energy is required to break an electron from the atom. The gap between the valance and conduction bands of an insulator is large. High energy is required to move an electron from the valance band to the conduction band. Therefore, an insulator subjected to a low voltage may have a small number of free electrons that are insufficient to create any current flow. However, if the same material is subject to a very large voltage, then there will be sufficient free electrons to initiate a current. Thus, an insulator that is used to cover a low voltage conductor of an electric cable may not be suitable for a high voltage cable.

III.2.3 Semiconductors

A semiconductor, in its intrinsic state, has the properties of neither a conductor nor an insulator. Silicon (Si) is the most commonly used semiconductor for power electronic devices. Intrinsic Si atoms form covalent bonds with their neighbouring atoms as shown in Figure III.2. As the temperature of the material increases, some of the covalent bonds break thus creating free electrons and holes. Both free electrons and holes contribute to the current flow.

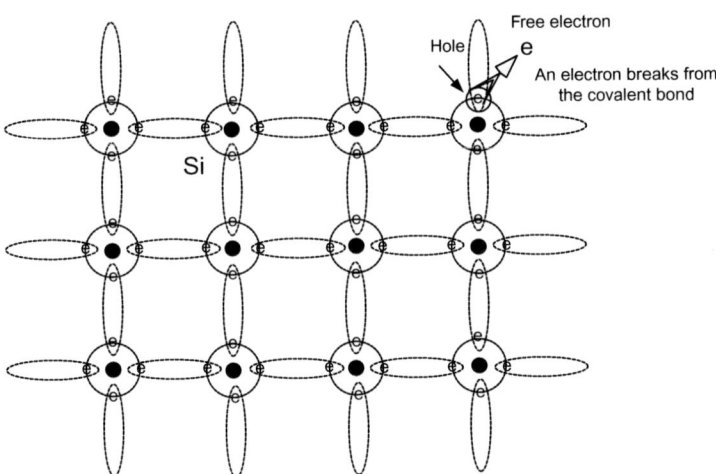

Figure III.2 Structure of an intrinsic semiconductor [1]

Figure III.3 shows the energy band structure of an intrinsic semiconductor. As an electron moves from an atom, the resultant hole remains at the valance band, while the free electron moves to the conduction band.

At room temperature, current flow in an intrinsic semiconductor is very small. In order to increase the conductivity of an intrinsic semiconductor, impurities are added, by doping. Two types of semiconductors are formed by doping: N-type semiconductor and P-type semiconductor.

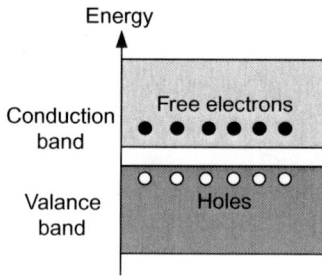

Figure III.3   Energy band of intrinsic semiconductor [1]

(a) N-type semiconductor

To form an N-type semiconductor a donor impurity with five outer electrons is added to the intrinsic semiconductor. Typically arsenic (As), phosphorous (P) or bismuth (Bi) is used as donor impurities. The 5th electron in the donor atom does not form a covalent bond and can easily be broken from the atom (Figure III.4), thus considered as a free electron.

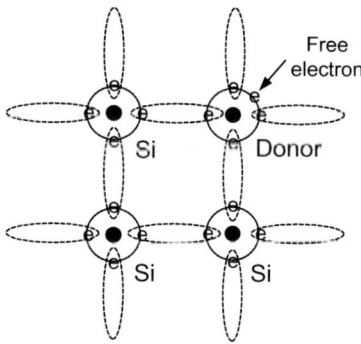

Figure III.4   Structure of N-type semiconductor [1]

(b) P-type semiconductor

To form a P-type semiconductor, an acceptor impurity with three outer electrons is added to the intrinsic Si. Boron (B), gallium (G) and indium (In) are commonly used for acceptor impurities. As shown in Figure III.5, a hole in the acceptor atom tries to attract an electron to form the missing covalent bond.

III.3   PN junction

By combining a P-type semiconductor with an N-type semiconductor a PN junction is formed. When P-type and N-type semiconductors are combined, due to the difference in the concentration of carriers (free electrons or holes) across the junction,

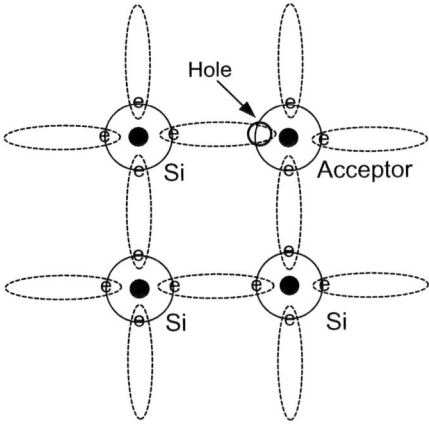

Figure III.5   Structure of P-type semiconductor [1]

free electrons from the N-type semiconductor cross the junction and combine with holes near the junction area. As shown in Figure III.6, this results in a negative ion as the neutrally charged atom has accepted an electron (negative charge). At the same time, holes from the P-type region cross the junction and combine with free electrons from the N-type semiconductor, thus producing positive ions. The resultant junction region will not have any charge carriers and is called the depletion region. As shown in Figure III.6, the depletion layer acts as a potential barrier to further movement of electrons and holes.

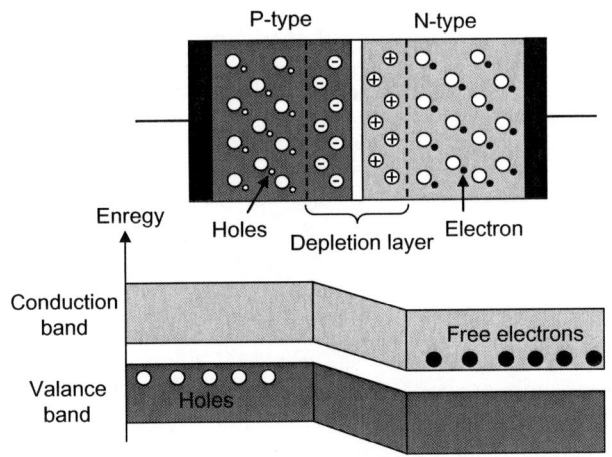

Figure III.6   A PN junction [1,2]

(a) Forward biased PN junction

Consider a case where a PN junction is connected to a battery (Figure III.7), where the P side is connected to the positive terminal of the battery and the N side is connected to the negative terminal of the battery. If the potential of the battery, E, is higher than the potential difference across the depletion layer, then the free electron on the N side and the holes on the P side will gain enough energy to overcome the potential barrier across the depletion layer, thus creating a current flow from the P side to the N side.

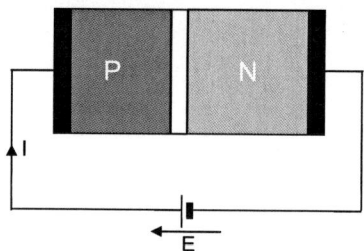

Figure III.7   Forward biased PN junction

(b) Reverse biased PN junction

If the positive terminal of the battery is connected to the N side of the PN junction, it will be reverse biased. In this case the electrons from the N side will be attracted towards the positive terminal of the battery, and the holes from the P side will be attracted towards the negative terminal of the battery. Therefore, the depletion layer gets broader thus preventing any current through the PN junction. However, there will be a very small current (leakage current) due to free electrons generated by thermal energy.

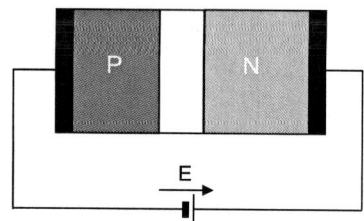

Figure III.8   Reverse biased PN junction

III.4   Diode

The device made from a single PN junction is called a diode. The symbol of a diode is shown in Figure III.9, where the direction of the arrow shows the direction of the flow of current. The P terminal is called the anode and the N terminal is called the cathode.

The V–I characteristic of a diode is shown in Figure III.10. The voltage V_γ is called the cut-in voltage, i.e. the voltage required to push electrons against the

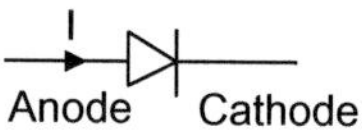

$$\xrightarrow{I} \quad \triangleright\!\!| $$
Anode   Cathode

Figure III.9   Symbol of a diode

potential difference of the depletion layer. The value of V_γ depends on the material used and is approximately 0.6 V for Si.

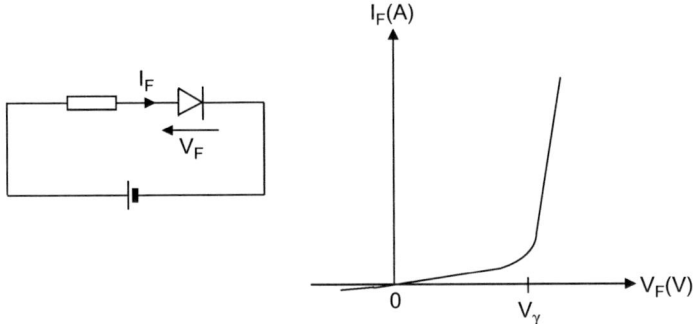

Figure III.10   V–I characteristic of a diode

EXAMPLE III.1

What is mean by rectification? Using suitable diagrams explain the operation of a single-phase rectifier.

Answer:

The conversion of an alternating current (bi-directional signal) into a direct current (uni-directional signal) is called rectification. As diodes let the current flow only in one direction, one of the main applications of a diode is rectification.

A half wave rectifier uses a single diode as shown in the following figure. During the positive half cycle of the AC voltage, the anode voltage is greater than the cathode voltage and the diode is in the forward biased region. During the negative half cycle the diode goes to the reverse biased region, thus blocking the AC input voltage.

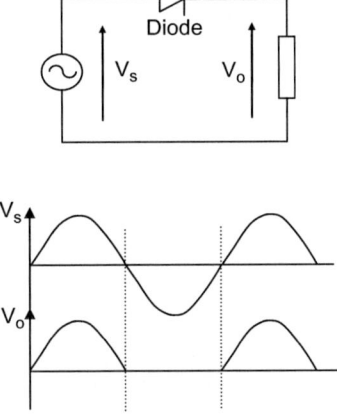

EXAMPLE III.2

Describe the operation of the full wave bridge rectifier shown in the following figure.

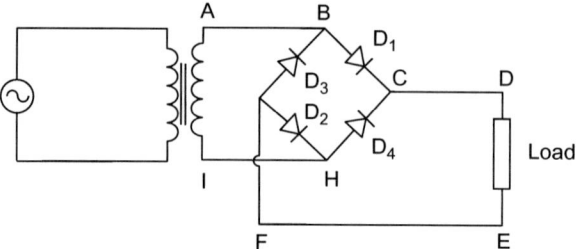

Answer:

During the positive half cycle, current flows in the path 'ABCDEFGHIA'. That is during this cycle, diodes D_1 and D_2 are forward biased and diodes D_3 and D_4 are reverse biased as shown in the following figure (top). During the negative half cycle, diodes D_3 and D_4 are forward biased and diodes D_1 and D_2 are reverse biased (see figure (bottom)) and current flows in the path 'IHCDEFGBAI'. That is during both positive and negative half cycles, the current through the load is in the same direction, thus giving a uni-directional voltage across the load.

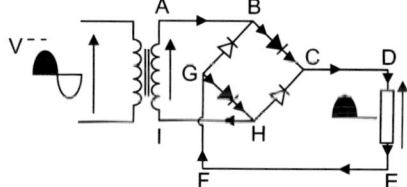

During the positive half cycle

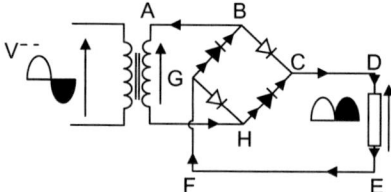

During the negative half cycle

III.5   Switching devices

The diode is an uncontrolled device that is when it is forward biased, current flows through it and there is no control over the current flow. In many power electronic applications controllable devices are required. In controllable devices, apart from the main current flowing terminals, a third terminal is used to control the device. Depending on the control techniques used, switching devices can be broadly categorised into: current-controlled devices and voltage-controlled devices.

III.5.1 Current-controlled devices

In a current-controlled device, the current through it is controlled by injecting a current into the third terminal. Commonly used current control devices are transistors, thyristors and gate turn OFF (GTO) thyristors. In transistors and GTO thyristors, the device can be turned ON and OFF using the control terminal, whereas in a thyristor the instance of turn ON can be controlled but the device will conduct until the direction of the current reverses, and at that point the device will naturally turn OFF (i.e. it will commutate).

1. Transistor [1,3,4]

A commonly used current-controlled device is the transistor. The transistor consists of two PN junctions formed by sandwiching P-N-P or N-P-N layers as shown in Figure III.11. In both types of transistor, the current through the collector and emitter terminals can be controlled by the base current. However, this control action only exists when the transistor is properly biased, that is by connecting appropriate voltages across the three terminals. Different biasing arrangements are employed for transistors. Figure III.12 shows a commonly used biasing arrangement, common-emitter biasing and the corresponding V–I characteristics. From the V–I characteristics, three operating regions can be recognised: active region, cut-off region and saturation region. In the active region, a small base current can control the collector current. As I_B increases, at some point the base current loses

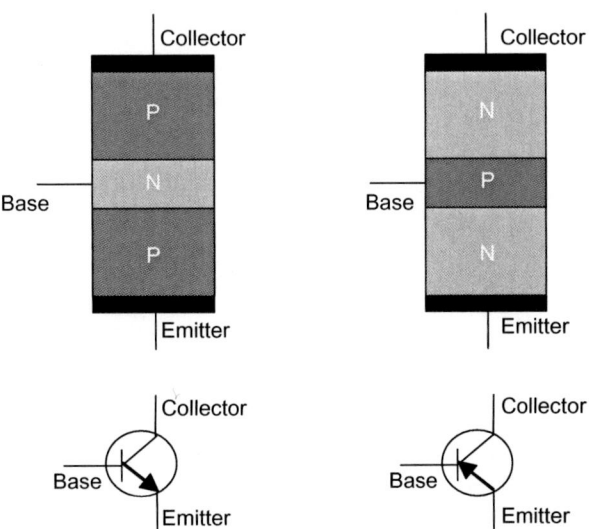

Figure III.11   Structure of a P-N-P and N-P-N transistor and their symbols

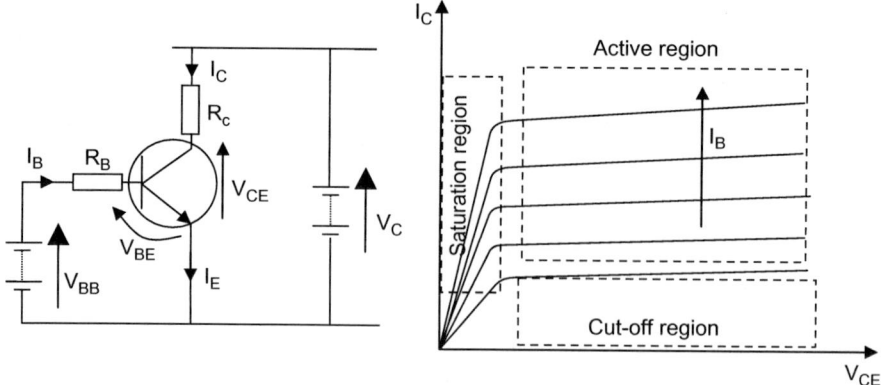

Figure III.12   Common-emitter biasing arrangement and V–I characteristics

control over the collector current. In this region (called the saturation region), the transistor acts as a closed switch. If I_B is reduced to a certain minimum, then I_C becomes zero and V_{CE} increases. This operation is equivalent to an open switch and this operating region is called cut-off region. In moderate or high power applications, transistors are not operated in the active region because of the rather high conduction losses of the device.

2. Thyristor [3,4]

The thyristor is a P-N-P-N device (Figure III.13) that can be turned ON by applying a gate pulse. Once the device is ON the gate loses its control and the device will naturally turn OFF when the current from the anode to cathode becomes very small. Thyristors are used in power applications such as induction motor drives and in very large power applications such as current source high voltage DC schemes. Thyristors with ratings up to 8.5 kV, 4000 A are currently available. As shown in Figure III.13, thyristor can be represented by two transistors, one a P-N-P and the other an N-P-N.

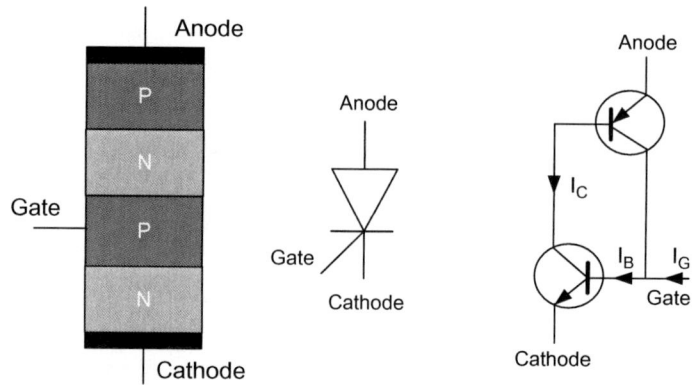

Figure III.13   Thyristor structure, symbol and transistor equivalent

The V–I characteristic of the thyristor is shown in Figure III.14. The characteristic can be divided into three regions and is explained using the two-transistor equivalent circuit:

(a) Reverse blocking region: Both transistors are in their cut-off regions and only a small leakage current flows through the device.
(b) Forward blocking region: The upper transistor is OFF and the lower one is ON. Thus there is no current through the device.
(c) Forward conduction region: When I_G is sufficiently large, the base current to the N-P-N transistor (I_B) is large enough to turn that transistor ON. This in turn increases the collector current, I_C, thus the base current of the upper P-N-P transistor. Then the collector current of the P-N-P transistor, that is the base current of the N-P-N transistor, increases. This cycle continues until both transistors go into saturation. Once both transistors are in the saturation region, the gate no longer has control over the current flow in the thyristor.

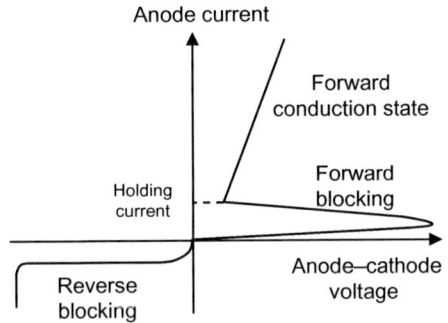

Figure III.14   V–I characteristic of the thyristor

3. Gate turn off (GTO) thyristor [3,4]

The GTO is a thyristor like P-N-P-N device having a complicated construction. It can be turned ON by a positive gate current and turned OFF by a negative gate current. However, a comparatively large current is required to turn it OFF. For example a 4000 A device may require a gate current as high as 750 A to turn it OFF. The symbol of the device is shown in Figure III.15.

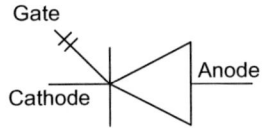

Figure III.15   Symbol of a GTO

III.5.2   Voltage-controlled devices

In voltage-controlled devices, the current through the device can be controlled by applying a suitable voltage to the control terminal. As the current drawn by the control terminal is very small these devices can directly be controlled using integrated circuits and microcontrollers.

1. MOSFET [1,3,4]

The field effect transistor (FET) has very similar applications to the transistor. The main difference is the way the device is turned ON and OFF. As the transistor is a current-controlled device, a small base current can control the collector current. On the other hand, in a FET a voltage applied across the gate and source can control the drain current. There are two types of FETs: junction field effect transistor (JFET) and metal oxide semiconductor field effect transistor (MOSFET). Only the MOSFET is treated in this text as they are generally used for medium power switching applications relevant to the scope of this book. Figure III.16 shows the typical structure and symbol of a MOSFET.

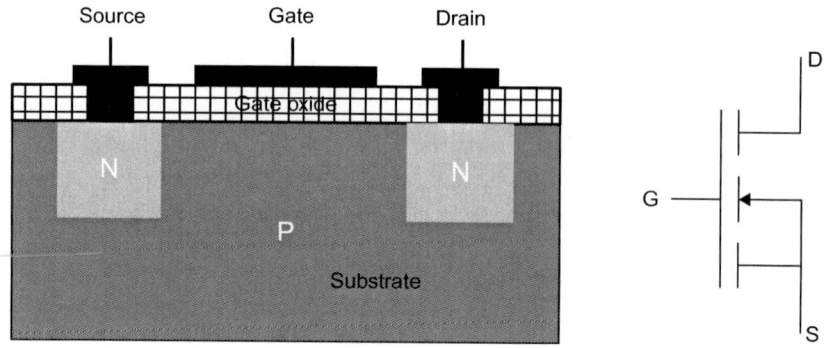

Figure III.16   MOSFET structure and symbol

There are two MOSFET types, namely, depletion type and enhancement type. Depending on the semiconductor type used for the substrate, each type of MOSFET is also categorised as N-type and P-type. Typical transfer characteristic of an N-type enhancement MOSFET is shown in Figure III.17. For a gate voltage of less than a threshold (typically 2 V), the device will not conduct. Once the gate voltage is increased beyond its threshold the device starts conducting and the drain current shows a quadratic characteristic with respect to the gate voltage.

2. IGBT [3,4]

The IGBT is a hybrid switch that consists of a MOSFET on its gate side and a transistor in its conduction path. Its equivalent circuit and symbol is shown in Figure III.18. One of the main disadvantages of a MOSFET is that its conduction losses are comparatively high when compared with that of a similar rated transistor. However, as a MOSFET is a voltage-controlled device that only draws very small gate current, it offers easy driving and lower losses in the driving circuits. Therefore, the IGBT offers the advantages of both transistors and MOSFETs. The IGBT has become a popular choice for medium power applications, and is now widely used for motor drives, wind power conversion systems and many other forms of distributed generation.

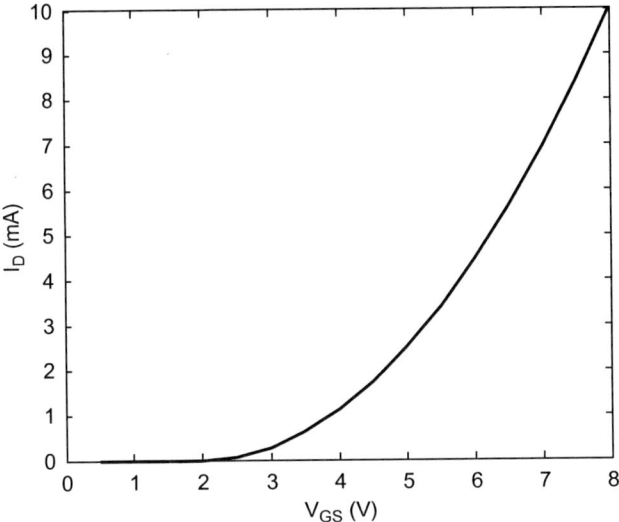

Figure III.17   Transfer characteristic of a MOSFET

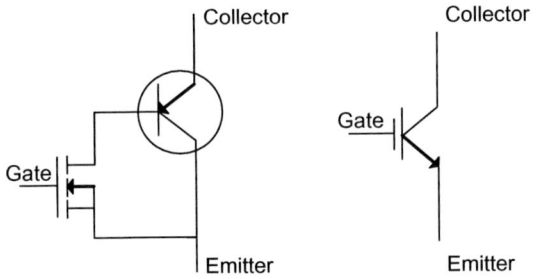

Figure III.18   IGBT symbol and equivalent circuit [3]

EXAMPLE III.3

The following figure shows a thyristor phase-controlled circuit. Draw the waveform of V_o and I_o when $\tan^{-1}(\omega L/R) \leq \alpha$, where α is the thyristor gating trigger angle.

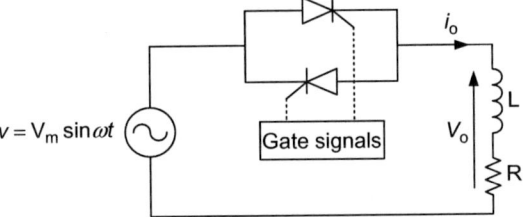

Answer:
If both thyristors are ON with a firing angle $\alpha = 0°$ (this is equivalent to L and R being directly connected to the source), then the current i_o is given by:

$$i_o = I_o \sin(\omega t + \phi), \quad \text{where} \quad \phi = \tan^{-1}\left(\frac{\omega L}{R}\right)$$

where I_o is the peak value of the current.

If $\alpha > \tan^{-1}(\omega L/R)$, then the thyristor will not conduct from ϕ to α as shown in the following figure. When the thyristors are not conducting, $v_o = 0$, otherwise v_o follows the input.

The waveforms of v_o and i_o are

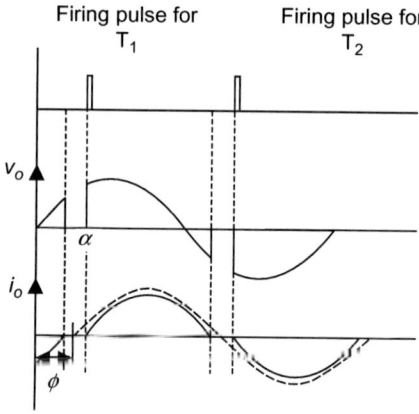

EXAMPLE III.4

A permanent magnet DC motor having an armature resistance of 2 Ω is connected to a 240 V battery via a DC chopper circuit (a transistor or a MOSFET switch which is operated in ON–OFF mode). On no load, the maximum speed possible is 1000 rev/min and then machine draws 1 A. For the DC motor induced voltage is proportional to the speed of rotation of the rotor.

1. Sketch a circuit diagram of the main components of the drive assuming the inductance of the motor is very large. Draw voltage and current waveforms when the chopper operates at 50% duty ratio.
2. Calculate the duty ratio of the switch to run the motor at 750 rev/min with full load current of 10 A.

Answer:
At 50% duty ratio, assuming inductance of the armature coil is much higher than its resistance:

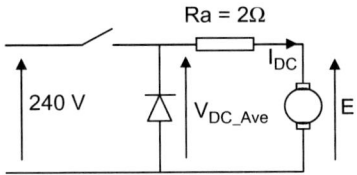

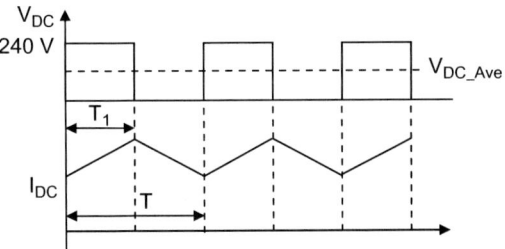

$V_{DC\_Ave} = d \times 240\,V$ where d is the duty ratio of the switch, which is given by T_1/T.

As the induced voltage is proportional to the speed, the maximum speed (1000 rev/min) is possible with the maximum armature voltage. Thus, chopper switch should operate with unity duty ratio.

$$V_{DC\_Ave} = 240\,V$$
$$E = 240 - 2 \times 1 = 238\,V$$

Since $E = kn$ (where n is the speed in revolutions per minute), at 1000 rev/min:

$$k = \frac{E}{n} = 0.238\,V/(rev/min)$$

Now when the machine runs at 750 rev/min:

$$E = 0.238 \times 750 = 178.5\,V$$
$$V_{DC\_Ave} = 178.5 + 2 \times 10 = 198.5\,V$$

The duty ratio of the switch = 198.5/240 = 0.83.

III.6   Voltage source inverters

A very large number of concepts and power electronic circuits have been used or proposed for the connection of small generators to the power system network. These have taken advantage of the operating characteristics of the various semiconductor switching devices that are available and, as improved switches are developed, then the circuits will continue to evolve. Some years ago, naturally commutated, current source, thyristor-based inverters were common. The thyristors are turned ON by a gate pulse but are turned OFF (or naturally commutated) by the external circuit. This type of equipment has the advantage of low electrical losses but the voltage of the DC link determines its power factor and its harmonic performance is poor. As well as the possible expense of filters, a converter with a poor harmonic performance requires time-consuming and expensive harmonic studies to ensure that its impact on the network will be acceptable. Such inverters are now rarely used in new distributed generation schemes.

Most modern converters use some form of voltage source converter (VSC), which, as the name implies, synthesise a waveform from a voltage source.

III.6.1   Single-phase voltage source inverter

A simple single-phase inverter is shown in Figure III.19. In this arrangement two voltage-controlled switches are used. For clarity they are shown as S_1 and S_2. These two switches are turned ON and OFF in a complementary manner so as to obtain a square waveform at the output of the inverter. When S_1 is ON and S_2 is OFF, the voltage across the load is $V_{DC}/2$. On the other hand when S_1 is OFF and S_2 is ON, the voltage across the load is $-V_{DC}/2$. Even though the output of this inverter is a square wave, from Fourier analysis it can be shown that the fundamental component of the voltage across the load is a sinusoid at the switching frequency of the switches.

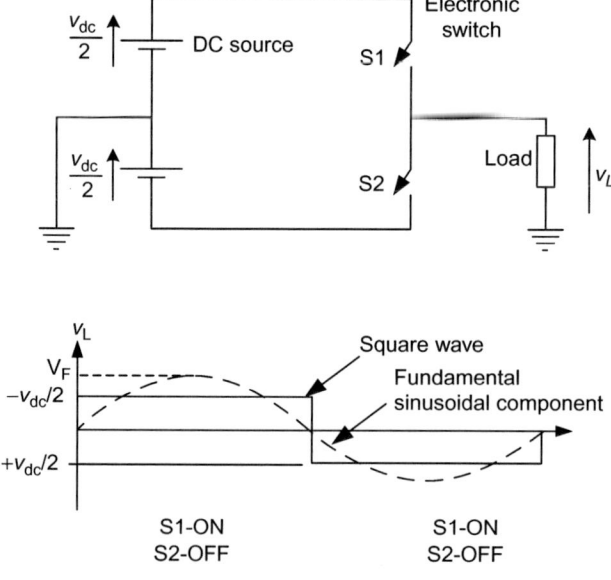

Figure III.19   A square wave inverter

1. Harmonics

From Fourier analysis, it can be shown that a square wave signal is formed by the addition of infinite number of sinusoids. The Fourier series of the square wave is given by:

$$v_L = \frac{4V_{DC}}{\pi} \sum_{n=1,3,5}^{\infty} \frac{\sin n\omega t}{n}$$

$$= V_F \sin \omega t + \frac{V_F}{3}\sin 3\omega t + \frac{V_F}{5}\sin 5\omega t + \frac{V_F}{7}\sin 7\omega t + \frac{V_F}{9}\sin 9\omega t...$$

$$(\text{III}.1)$$

where $V_F = 4V_{DC}/\pi$ is the peak value of the fundamental (shown in dotted in Figure III.19).

As the output contains significant higher order frequency components (harmonics), this inverter has limited applications (and is mainly used for low-cost off-grid systems). The harmonic voltages increase the losses in the loads connected to the inverter and create pulsating torques in any connected motors and generators. The utilities also impose restrictions on the harmonics that can be generated by power electronic equipment to minimise the nuisance caused by harmonics to the other consumers connected in the utility grid.

In order to minimise the harmonic generated by the converter, that is to synthesise an approximation to a sine wave from a DC voltage source, various modulation strategies are used and they include the following.

- Carrier modulated techniques that compare a reference signal with a trigger signal. The most well known of these is sinusoidal pulse width modulation (PWM) which is easily implemented in hardware.
- Hysteresis control.
- Programmed pulse width modulation (sometimes known as selective harmonic elimination, SHEM).
- Space vector modulation.

In the case of pulse width modulation (PWM) technique two switches, S_1 and S_2, are turned ON and OFF as shown in Figure III.20. A modulation signal at the same frequency as the fundamental required at the inverter output is compared with a high frequency triangular carrier signal. Typically, a carrier of 2–30 kHz is used. When the triangular signal is greater than the modulation signal switch S_1 is turned ON and switch S_2 is turned OFF in a complementary manner. The resultant output voltage waveform of this inverter is shown in the bottom trace of Figure III.20.

In hysteresis control the output is controlled to within reference bands on either side of the desired wave. In programmed pulse width modulation sophisticated optimisation strategies are used, off-line, to determine the required switching angles to eliminate particular harmonics. In space vector modulation the output voltage space vector of the converter is defined and synthesised. This is convenient to implement using a digital controller through a two-axis transformation.

Even though the output voltage has a number of pulses, from Fourier analysis it can be shown that the output of the PWM inverter has reduced lower order harmonics to less than that of a square wave inverter.

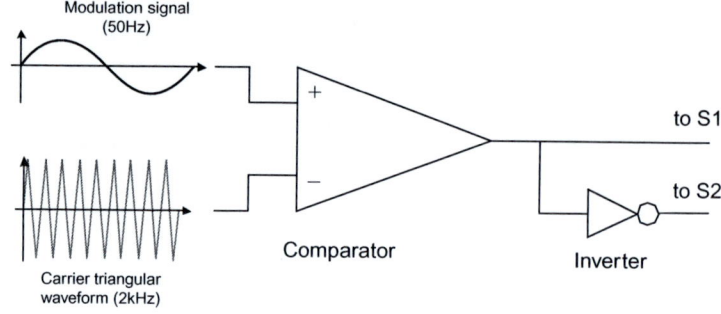

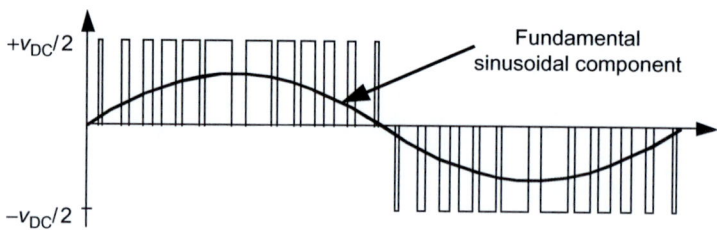

Figure III.20   Pulse width modulation technique

2. Losses

Each pulse of the PWM waveform has two transients, turn ON transient and turn OFF transient, shown in Figure III.21 (for clarity the transient times are exaggerated in the figure).

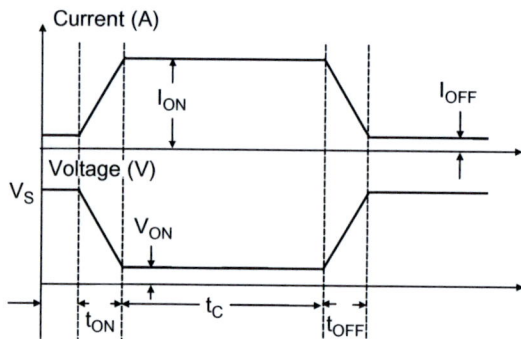

Figure III.21   Typical turn ON and turn OFF transients of switching devices

The total power dissipation is the sum of the switching transition losses (turn ON and turn OFF) and the ON state conduction loss. For the switching transitions shown in Figure III.21, the total loss is approximately given by (III.2) [3]:

$$\text{Total losses} = \left[\frac{V_S I_{ON} t_{ON}}{6} + \frac{V_S I_{ON} t_{OFF}}{6} + V_{ON} I_{ON} t_C \right] \times f_s \qquad \text{(III.2)}$$

where f_s is the switching frequency (frequency of the carrier signal) of the PWM signal.

From (III.2), it is clear that the total loss increases with the switching frequency.

III.6.2 Three-phase voltage source inverter

In many applications three-phase inverters are utilised. The operation of a three-phase inverter is similar to the operation of three single-phase inverters, each producing an output phase shifted by 120° to the other. Figure III.22 shows a three-phase inverter having six MOSFETs. The MOSFET pairs S_{a1} and S_{a2}, S_{b1} and S_{b2}, and S_{c1} and S_{c2} are controlled in complementary manner.

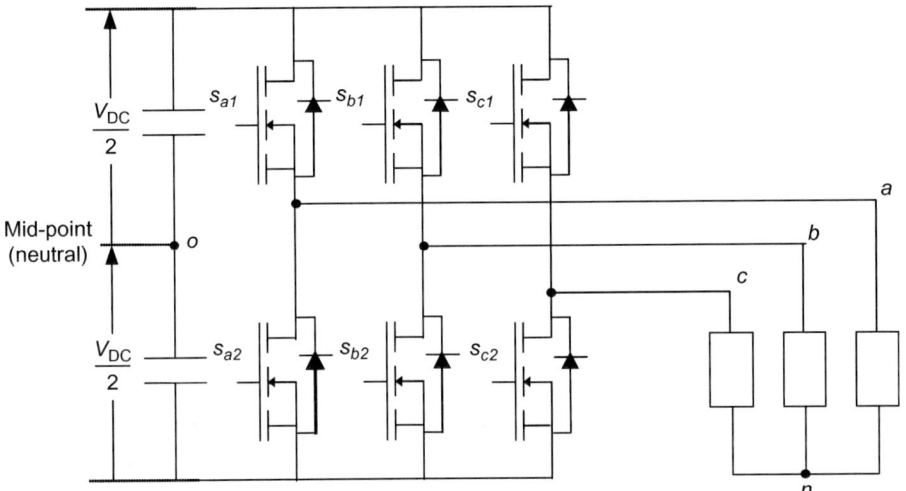

Figure III.22 MOSFET-based three-phase inverter

III.7 Problems

1. Discuss the difference between the insulator, conductor and semiconductor using energy bands.

    (*Answer*: See Section III.2)

2. A 230 V AC mains is connected to a full-bridge diode rectifier through a 5:1 step down transformer. The ON-state drop across each diode is 0.6 V. What is the peak value of the output of the bridge when it is not connected to a load?

    (*Answer*: 230/5 − (2 × 0.6) = 45.8 V)

3. A thyristor phase-controlled circuit is used to control the current through a load of 10 Ω resistance and 4 Ω reactance. If the thyristor firing angle is 30°, draw the waveform of the current.

   (*Answer*: $\alpha = 30° > \tan^{-1}(4/10)$ – see Example III.5.3 (3) for the plot)

4. A three-phase square wave inverter is connected to a DC source of 100 V. Calculate the magnitude of three lowest order harmonics produced by the inverter.

   (*Answer*: Using (III.1): 5th = 25.5 V, 7th = 18.2 V and 11th = 11.6 V)

5. The collector of an IGBT is connected to a 300 V supply through a 10 Ω resistor. The IGBT is turned ON and OFF at 10 kHz. Turn ON and turn OFF times of the IGBT are 50 ns and 400 ns respectively. The turn ON voltage drop can be neglected. Calculate the total losses associated with the IGBT.

   (*Answer*: Using (III.2): Total losses = 6.75 W)

III.8   Further reading

1. Milliman J., Halkias C.C. *Integrated Electronics: Analog and Digital Circuits and Systems*. Mc Graw-Hill; 1972, ISBN 0-070-423156.
2. Markvart T. *Solar Electricity*. Wiley; 1994, ISBN 0-471-941611.
3. Williams B.W. *Power Electronics: Devices, Drives, Applications and Passive Components*. MacMillan Press; 1992, ISBN 0-333-57351X.
4. Mohan N., Undeland T.M., Robbins W.P. *Power Electronics – Converters, Applications and Design*. 2nd edn. New York: John Wiley & Sons, Inc.; 1995, ISBN 0-471-58408-8.

Tutorial IV
Power systems

Lindsey Oil Refinery Co-generation Plant
The plant produces 118 MW heat and 38 MW of electrical energy [National
Power PLC]

IV.1   Introduction

The power system converts mechanical energy into electrical energy using generators, then transmits the electricity over long distances and finally distributes it to domestic, industrial and commercial loads. Generation is at a low voltage (400 V to around 25 kV) and then the voltage is stepped up to transmission voltage levels (e.g. 765 kV, 400 kV, 275 kV) and finally stepped down to distribution voltages (e.g. 13.8 kV, 11 kV or 400 V). Each of these conversion stages takes place at a substation with a number of different pieces of equipment to: (a) transform the system voltage (power transformers), (b) break the current during faults (circuit breakers), (c) isolate a section for maintenance (isolators) after breaking the current, (d) protect the circuit against lightning overvoltages (surge arresters) and (e) take voltage and current measurements (voltage transformers – VT and current transformers – CT). In addition to this primary plant, which carries the main current, secondary electronic equipment is used to monitor and control the power system as well as to detect faults (short-circuits) and control the circuit breakers.

Practical AC power systems use three phases that are of the same magnitude and displaced 120° degrees electrical from each other as discussed in Tutorial I. When the three phases are thus balanced, no current flows in the neutral and so at higher voltages only three phase conductors are used and the neutral wire is omitted. In order to represent the power system in control diagrams and reports, a single line representation is used; the three-phase lines are shown by a single line. Typical single line diagram symbols used for a balanced power system arc given in Table IV.1 and a single line diagram is shown in Figure IV.1.

IV.2   Power transformers

The power transformer is a key component of the power system, Figure IV.1, increasing the voltage from the generators and then reducing it for the loads.

One coil is connected to a voltage source (directly or through other power system components) and called the primary winding (N_1) and the other coil is connected to a load and called the secondary winding (N_2). A mutual magnetic flux produced by the alternating current in the primary winding links with the secondary winding and induces a voltage in it. The equivalent circuit of an ideal transformer, where flux produced by the primary links completely with the secondary, is shown in Figure IV.2(b).

If flux in the core is $\phi = \phi_m \sin \omega t$, then from Faraday's law:

$$E_1 = N_1 \frac{d\phi}{dt} = \omega N_1 \phi_m \cos \omega t \qquad (IV.1)$$

where E_1 is the internal voltage of the primary.

Assuming that ϕ links completely with the secondary of the transformer, the secondary internal voltage:

$$E_2 = N_2 \frac{d\phi}{dt} = \omega N_2 \phi_m \cos \omega t \qquad (IV.2)$$

Table IV.1   Typical symbols used for one line diagrams

Symbol	Description
	Generator
	Motor
	Transmission line or cable
	Busbar
	Transformer (two winding) Upper symbol – commonly used in the Europe Lower symbol – commonly used in the United States
	Transformer with tap changer
	Resistor
	Reactor
	Capacitor
	Circuit breaker

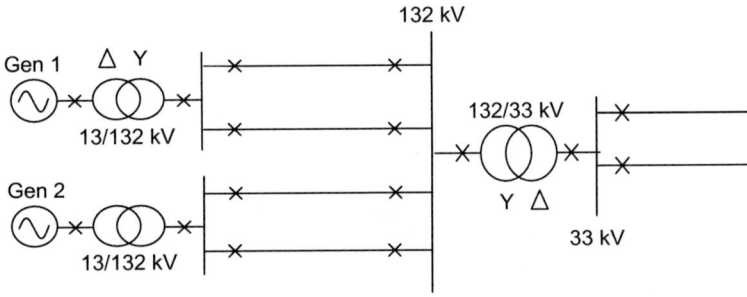

Figure IV.1   One line diagram of part of the power system

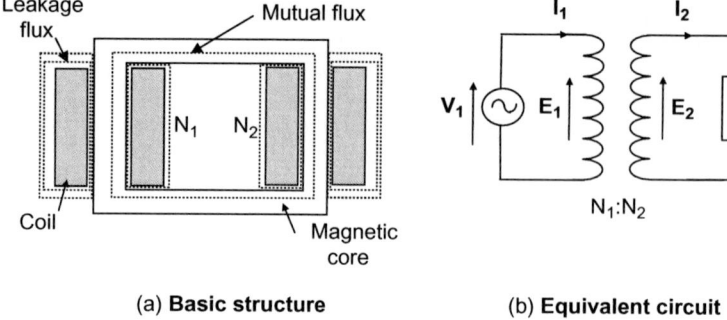

(a) **Basic structure**   (b) **Equivalent circuit**

Figure IV.2   A single-phase transformer

Dividing (IV.1) by (IV.2), the following relationship is obtained for the ratio of the magnitudes of E_1 and E_2:

$$\frac{E_1}{E_2} = \frac{N_1}{N_2} \tag{IV.3}$$

Equating the volt-ampere rating of the primary and secondary:

$$E_1 I_1 = E_2 I_2 = \frac{N_2}{N_1} E_1 I_2$$

$$\frac{I_1}{I_2} = \frac{N_2}{N_1} \tag{IV.4}$$

In a real transformer, a small amount of flux links only with the primary and that component of the flux is called the leakage flux. A similar leakage flux exists on the secondary side.

For analysis of a transformer, it is convenient to use an equivalent circuit. Figure IV.3 shows the equivalent circuit of the transformer where R_1 and R_2 represent the resistances of the primary and secondary windings, X_1 and X_2 represent the reactances of the windings due to leakage fluxes, X_m represents the reactance of the mutual flux, R_c represents the core losses, and E_1 and E_2 are the internal voltage on each coil.

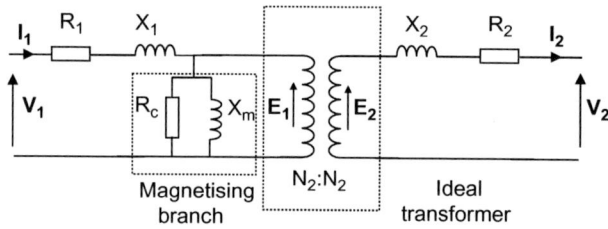

Figure IV.3   A basic structure of a transformer [1]

A resistor in one side (primary or secondary) of the transformer circuit can be referred to the other side by equating the energy dissipated in them as heat. For example, the heat loss associated with R_1 (on the primary side) is $I_1^2 R_1$. If that resistor is transferred to the secondary side an equivalent resistance is R'_1, then:

$$I_2^2 R'_1 = I_1^2 R_1$$
$$R'_1 = \left(\frac{I_1}{I_2}\right)^2 R_1 \qquad (IV.5)$$

Substituting for the current ratio from (IV.4) into (IV.5):

$$R'_1 = \left[\frac{N_2}{N_1}\right]^2 R_1 \qquad (IV.6)$$

The primary leakage reactance may also be referred to the secondary side in a similar manner:

$$X'_1 = \left[\frac{N_2}{N_1}\right]^2 X_1 \qquad (IV.7)$$

Typically, the current through the magnetising branch of a transformer (X_m and R_c) is only about 3–5% of the full-load current. Therefore when considering a loaded transformer, the magnetising branch is usually omitted. The equivalent circuit of a loaded transformer where all the quantities are referred to the secondary is shown in Figure IV.4.

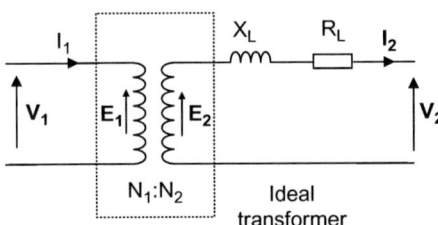

Figure IV.4   Simplified circuit of a transformer

In Figure IV.4:

$$R_L = R_2 + \left[\frac{N_2}{N_1}\right]^2 R_1$$
$$X_L = X_2 + \left[\frac{N_2}{N_1}\right]^2 X_1 \qquad (IV.8)$$

The windings of practical power transformers are usually tapped to allow alteration of the number of turns in use and hence the turns ratio of the transformer.

In this way the secondary voltage may be altered (with a constant primary voltage). On load tap changers use a combination of diverter and selector switches to change the transformer ratio while it passes current and is on load (Figure IV.5). Off load tap changers can be operated with voltage on the transformer (but no current flowing), while off circuit tap changers can only be operated when the transformer is isolated.

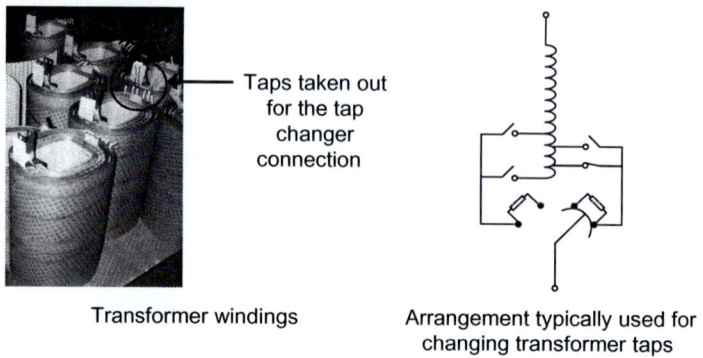

Taps taken out for the tap changer connection

Transformer windings

Arrangement typically used for changing transformer taps

Figure IV.5   Transformer winding and tap changing arrangement

EXAMPLE IV.1

A 50 kVA transformer has 400 turns on the primary and 40 turns on the secondary. The primary and secondary resistances are 0.3 Ω and 0.01 Ω respectively, and the corresponding leakage reactances are $j1.1$ Ω and $j0.035$ Ω respectively. Calculate the equivalent impedance referred to the primary.

Answer:

Since the resistance and reactance are transformed from the secondary to the primary, the transformation is based on the square of the primary to secondary turns ratio, thus if the equivalent impedance referred to the primary is $R_{Lp} + jX_{Lp}$:

$$R_{Lp} = R_1 + \left[\frac{N_1}{N_2}\right]^2 R_2$$

$$= 0.3 + \left[\frac{400}{40}\right]^2 \times 0.01$$

$$= 1.3\,\Omega$$

$$X_{Lp} = X_1 + \left[\frac{N_1}{N_2}\right]^2 X_2$$

$$= 1.1 + \left[\frac{400}{40}\right]^2 \times 0.035$$

$$= 4.6\,\Omega$$

IV.3   Per-unit system

The power system has multiple voltage levels from 765 kV, down to 400 V or even 120 V and this makes circuit analysis rather confusing. By converting all the quantities into dimensionless quantities, known as per-unit (pu) quantities, it is much easier to analyse the power system. The pu quantity of any value is defined as the ratio between the actual value of the quantity and the base or reference value in the same unit. The per-unit system avoids confusion with transformers, where the impedance of transformers depends on the side from which it is viewed, and eliminates the factor of $\sqrt{3}$ involved in line and phase voltage and current quantities. The per-unit normalisation applies only to magnitudes and angles are not normalised.

The per-unit calculation starts with defining base values for power system quantities. First, choose one base power (VA) for the entire power system being studied. This value is arbitrarily chosen, normally appropriate for the size of the system (a common choice for a large power system is 100 MVA). Second, one base voltage is chosen for each voltage level. Usually this is the nominal voltage on each side of the transformers; defined by the nominal turns ratio of the transformer. Finally, other base values are calculated to get the same relationship between per-unit quantities as between actual quantities.

For a distributed generator calculation, a suitable MVA base, S_b, is first chosen. The line-to-line voltages, V_L, of the various voltage levels are selected as the voltage bases (V_b). The current and impedance base are then obtained as:

$$I_b = \frac{S_b}{\sqrt{3}V_b} \tag{IV.9}$$

$$Z_b = \frac{V_b/\sqrt{3}}{I_b} = \frac{V_b/\sqrt{3}}{S_b/\sqrt{3}V_b} = \frac{[V_b]^2}{S_b} \tag{IV.10}$$

For example, if $V_L = 33$ kV and $S_b = 100$ MVA, Z_b is given by:

$$Z_b = \frac{(33 \times 10^3)^2}{100 \times 10^6} = 10.86\,\Omega$$

IV.3.1   Power transformers in per unit

For three-phase transformers in a power system, the resistances are much smaller than reactances. Therefore, the per-phase transformer equivalent circuit shown in Figure IV.4 is simplified by neglecting the resistance (Figure IV.6).

For a transformer two base voltages are defined: primary voltage, V_{L1}, and the secondary voltage, V_{L2}.

From Figure IV.6:

$$E_1 = V_1 \tag{IV.11}$$

$$E_2 = V_2 + jI_2X_L \tag{IV.12}$$

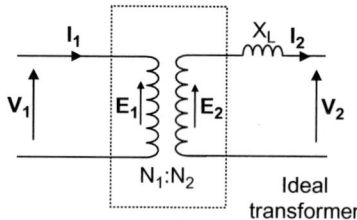

Figure IV.6   A simplified circuit of a power transformer

By dividing (IV.11) by $V_{L1}/\sqrt{3}$ (as the equation is a per-phase equation):

$$\mathbf{E_{1,pu}} = \mathbf{V_{1,pu}} = 1 \text{ pu} \tag{IV.13}$$

Similarly by dividing (IV.12) by $V_{L2}/\sqrt{3}$:

$$\mathbf{E_{2,pu}} = \mathbf{V_{2,pu}} + \frac{j\mathbf{I_2}X_L}{(V_{L2}/\sqrt{3})} = 1 \text{ pu} \tag{IV.14}$$

From (IV.9) and (IV.10):

$$I_b Z_b = V_{L2}/\sqrt{3} \tag{IV.15}$$

Therefore, (IV.14) was rewritten as:

$$\mathbf{E_{2,pu}} = \mathbf{V_{2,pu}} + j\mathbf{I_{2,pu}}X_{L,pu} = 1 \text{ pu} \tag{IV.16}$$

From (IV.13) and (IV.16):

$$\mathbf{V_{1,pu}} = \mathbf{V_{2,pu}} + j\mathbf{I_{2,pu}}X_{L,pu} \tag{IV.17}$$

Therefore, the transformer is represented by the equivalent circuit shown in Figure IV.7 in per unit.

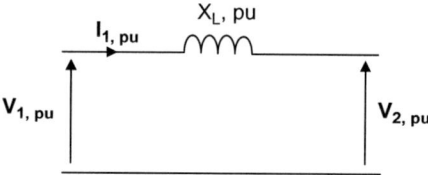

Figure IV.7   Transformer equivalent circuit in pu

The equivalent leakage reactance of the transformer, X_L, is normally defined as a percentage on its own name-plate rating. For example a transformer impedance of 6% on a 1 MVA transformer is simply: $X_L = 0.06$ pu, $S_b = 1$ MVA.

IV.3.2 Generators

As discussed in Tutorial II, a generator also has an internal impedance where the resistive component is very small compared to the inductive component. The internal impedance of the generator is generally represented by a percentage, which is the internal reactance in pu on the machine base multiplied by 100.

IV.3.3 System studies

To study a system using the pu system, all quantities are expressed in a consistent manner, that is on the same base power for the whole system being studied and the same base voltage for all the components at a voltage level. In many cases, the impedances of the transformers and generators are given on their own name-plate VA bases. It is then necessary to convert these impedances into the common system base.

Assume that a pu impedance, Z_1, is given on a power base of S_{b1}. If the ohmic value of the impedance is Z. Then from (IV.10):

$$Z_1 = \frac{Z}{Z_{b1}} = Z \times \frac{S_{b1}}{V_L^2} \tag{IV.18}$$

The impedance Z can be translated to a per unit impedance, Z_2, on a power base of S_{b2} as:

$$Z_2 = \frac{Z}{Z_{b2}} = Z \times \frac{S_{b2}}{V_L^2} \tag{IV.19}$$

From (IV.18) and (IV.19):

$$Z_2 = Z_1 \times \frac{S_{b2}}{S_{b1}} \tag{IV.20}$$

EXAMPLE IV.2

Using a 10 MVA base, change all the parameters in the following circuit into per unit and draw the simplified equivalent circuit.

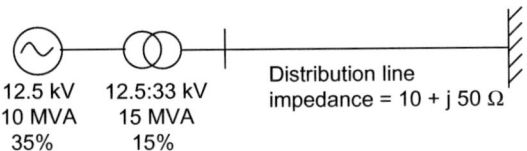

12.5 kV   12.5:33 kV   Distribution line
10 MVA   15 MVA   impedance = 10 + j 50 Ω
35%   15%

Answer:
Assume voltage bases of: 12.5 kV for the primary side of the transformer and 33 kV for the secondary side of the transformer.

On 10 MVA base, the generator reactance is 35% = j 0.35 pu.

The transformer impedance is equal to j 0.15 pu on a 15 MVA base. From (IV.20), on a 10 MVA base it is equal to j 0.15 $\times$ 10/15 = j 0.1 pu.

On 33 kV, 10 MVA, the base impedance is given by: $Z_b = (33 \times 10^3)^2/ 10 \times 10^6 = 108.6\ \Omega$

The per unit impedance of the distribution line is $(10 + j50)/108.6 = 0.092 + j0.46$ pu.

Then system in pu is drawn as:

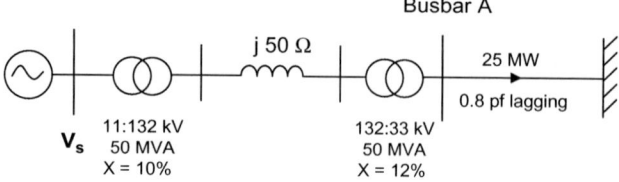

i 0.35   j 0.1

1 pu   0.092 + j 0.46

EXAMPLE IV.3

A schematic diagram of a radial network with a distributed generator is shown here. Calculate the terminal voltage of the synchronous generator if the voltage at busbar A at 30 kV. Use a 100 MVA base.

Busbar A

j 50 Ω

25 MW

0.8 pf lagging

11:132 kV   132:33 kV
V_s 50 MVA   50 MVA
X = 10%   X = 12%

Answer:

On $V_b = 132$ kV and $S_b = 100$ MVA, Z_b is given by:

$$Z_b = \frac{(132 \times 10^3)^2}{100 \times 10^6} = 174.24\ \Omega$$

The 11:132 kV transformer impedance on 100 MVA base $= j0.1 \times (100/50) = j0.2$ pu.

The 132:33 kV transformer impedance on 100 MVA base $= j0.12 \times (100/50) = j0.24$ pu.

The line impedance (j 50 Ω) on 132 kV, 100 MVA base $= j50/174.24 = j0.287$ pu.

Since $\cos \phi = 0.8$, $\phi = 36.87°$

Therefore, reactive power absorbed by the main system $= (25/\cos \phi) \times \sin \phi = (25/0.8) \times 0.6. = 18.75$ MVAr

Load power in pu $= (P + jQ)/S_b = (25 + j18.75)/100 = 0.25 + j0.1875$ pu.

As the voltage at busbar A is 30 kV, on a 33 kV base it is 30/33 pu $= 0.909$ pu. Defining this voltage as the reference voltage:

$$V_L I_L^* = 0.25 + j0.1875$$

$$0.909 \times I_L^* = 0.25 + j0.1875$$

$$I_L = 0.275 - j0.206$$

Now for the radial distribution system, the following equation was written in pu:

$$\mathbf{V_S} = V_L + \mathbf{I_L} \times j(0.2 + 0.24 + 0.287)$$
$$= 0.909 + (0.275 - j0.206) \times j0.727$$
$$= 0.909 + 0.15 + j0.2$$
$$= 1.059 + j0.2$$
$$= 1.08\angle 10.7° \text{ pu}$$

On the 11 kV base, the generator terminal voltage = $1.08 \times 11 = 11.88$ kV.

IV.4   Symmetrical components

A three-phase system operating under balanced, symmetrical conditions was discussed in Tutorial I. However, unsymmetrical conditions often occur in the power system due to unbalanced loading and asymmetrical faults (line–ground, line–line, etc.). In order to analyse the power system under unsymmetrical conditions, the system of symmetrical components is used. The basis is that any set of unbalanced three-phase voltages or currents can be represented by three balanced sets of phasors – positive, negative and zero – as shown in Figure IV.8.

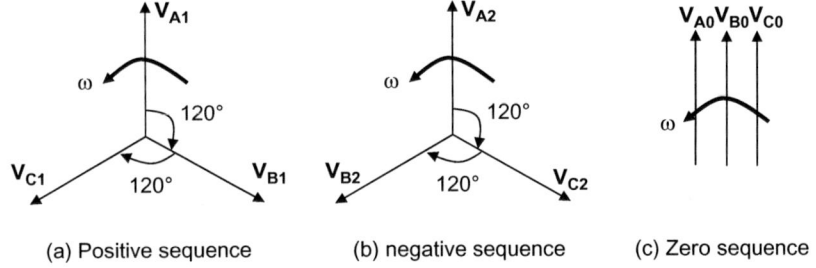

(a) Positive sequence   (b) negative sequence   (c) Zero sequence

Figure IV.8   Phasor representation of a three-phase voltage system

The positive sequence phasor voltages have the same phase sequence[1] (ABC) as the three-phase electrical power system. If the phase sequence of three voltages is ACB, then these voltage components are called the negative sequence components. The components where three voltages are equal in magnitude and phase are called zero sequence components.

The phasor representation of the zero sequence voltages (subscript 0 refers to zero sequence):

$$\mathbf{V_{A0}} = \mathbf{V_{B0}} = \mathbf{V_{C0}} = V_0\angle 0° = V_0 \times e^{j0} \qquad \text{(IV.21)}$$

---

[1] See Figure I.13 in Tutorial I.

The phasor representation of the positive sequence voltages (subscript 1 refers to positive sequence):

$$\mathbf{V_{A1}} = V_1 \angle 0° = V_1 \times e^{j0}$$
$$\mathbf{V_{B1}} = V_1 \angle -120° = V_1 \times e^{-j2\pi/3} = V_1 \times e^{j4\pi/3}$$
$$\mathbf{V_{C1}} = V_1 \angle -240° = V_1 \times e^{-j4\pi/3} = V_1 \times e^{j2\pi/3}$$
(IV.22)

Let $\lambda = e^{j2\pi/3}$, then:

$$\mathbf{V_{B1}} = \mathbf{V_{A1}} \times \lambda^2$$
$$\mathbf{V_{C1}} = \mathbf{V_{A1}} \times \lambda$$
(IV.23)

The phasor representation of the negative sequence voltages (subscript 2 refers to the negative sequence):

$$\mathbf{V_{A2}} = V_2 \angle 0° = V_2 \times e^{j0}$$
$$\mathbf{V_{B2}} = V_2 \angle -240° = V_2 \times e^{-j4\pi/3} = V_2 \times e^{j2\pi/3}$$
$$\mathbf{V_{C2}} = V_2 \angle -120° = V_2 \times e^{-j2\pi/3} = V_2 \times e^{j4\pi/3}$$
(IV.24)

Then:

$$\mathbf{V_{B2}} = \mathbf{V_{A2}} \times \lambda$$
$$\mathbf{V_{C2}} = \mathbf{V_{A2}} \times \lambda^2$$
(IV.25)

EXAMPLE IV.4

Using a phasor diagram show that the three unbalanced currents $\mathbf{I_A} = 200 \angle 10°$, $\mathbf{I_B} = 250 \angle -100°$ and $\mathbf{I_C} = 150 \angle -200°$ can be represented by the addition of positive, negative and zero sequence currents.

Answer:

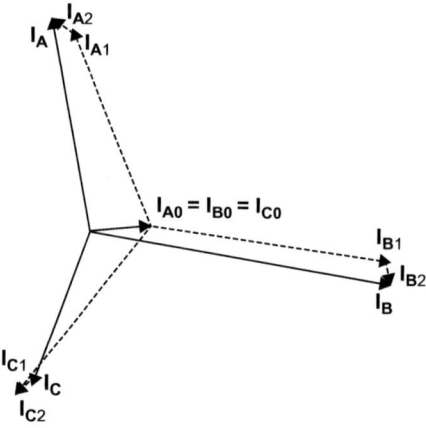

As shown in the above diagram, the unbalanced currents can be represented by the addition of positive sequence currents (dashed lines) having a magnitude of 196 A, where $\mathbf{I}_{A1}$ leads $\mathbf{I}_A$ by 22°, negative sequence currents (dotted lines) having a magnitude of 19 A, where $\mathbf{I}_{A2}$ leads $\mathbf{I}_A$ by 57°, and zero sequence currents (solid lines) having a magnitude of 54 A, where $\mathbf{I}_{A0}$ lags $\mathbf{I}_A$ by 85°.

As any unbalanced set of three-phase voltages or currents can be represented by the addition of the zero, positive and negative sequence voltage or currents, for voltages:

$$\begin{aligned}
\mathbf{V}_A &= (\mathbf{V}_{A0} + \mathbf{V}_{A1} + \mathbf{V}_{A2}) \\
\mathbf{V}_B &= (\mathbf{V}_{B0} + \mathbf{V}_{B1} + \mathbf{V}_{B2}) \\
\mathbf{V}_C &= (\mathbf{V}_{C0} + \mathbf{V}_{C1} + \mathbf{V}_{C2})
\end{aligned} \tag{IV.26}$$

Equation (IV.26) was rewritten by substituting from (IV.21), (IV.23) and (IV.25):

$$\begin{aligned}
\mathbf{V}_A &= (\mathbf{V}_{A0} + \mathbf{V}_{A1} + \mathbf{V}_{A2}) \\
\mathbf{V}_B &= (\mathbf{V}_{A0} + \lambda^2\mathbf{V}_{A1} + \lambda\mathbf{V}_{A2}) \\
\mathbf{V}_C &= (\mathbf{V}_{A0} + \lambda\mathbf{V}_{A1} + \lambda^2\mathbf{V}_{A2})
\end{aligned} \tag{IV.27}$$

In matrix form:

$$\begin{bmatrix} \mathbf{V}_A \\ \mathbf{V}_B \\ \mathbf{V}_C \end{bmatrix} = \begin{bmatrix} 1 & 1 & 1 \\ 1 & \lambda^2 & \lambda \\ 1 & \lambda & \lambda^2 \end{bmatrix} \begin{bmatrix} \mathbf{V}_{A0} \\ \mathbf{V}_{A1} \\ \mathbf{V}_{A2} \end{bmatrix} \tag{IV.28}$$

If $\mathbf{V}_A$, $\mathbf{V}_B$ and $\mathbf{V}_C$ are known then the three sequence component may be found by:

$$\begin{bmatrix} \mathbf{V}_{A0} \\ \mathbf{V}_{A1} \\ \mathbf{V}_{A2} \end{bmatrix} = \begin{bmatrix} 1 & 1 & 1 \\ 1 & \lambda^2 & \lambda \\ 1 & \lambda & \lambda^2 \end{bmatrix}^{-1} \begin{bmatrix} \mathbf{V}_A \\ \mathbf{V}_B \\ \mathbf{V}_C \end{bmatrix} \tag{IV.29}$$

In order to find the inverse of

$$\begin{bmatrix} 1 & 1 & 1 \\ 1 & \lambda^2 & \lambda \\ 1 & \lambda & \lambda^2 \end{bmatrix}, \begin{bmatrix} 1 & 1 & 1 \\ 1 & \lambda^2 & \lambda \\ 1 & \lambda & \lambda^2 \end{bmatrix} \times \begin{bmatrix} 1 & 1 & 1 \\ 1 & \lambda & \lambda^2 \\ 1 & \lambda^2 & \lambda \end{bmatrix}$$

is considered:

$$\begin{bmatrix} 1 & 1 & 1 \\ 1 & \lambda^2 & \lambda \\ 1 & \lambda & \lambda^2 \end{bmatrix} \begin{bmatrix} 1 & 1 & 1 \\ 1 & \lambda & \lambda^2 \\ 1 & \lambda^2 & \lambda \end{bmatrix} = \begin{bmatrix} 3 & (1+\lambda+\lambda^2) & (1+\lambda+\lambda^2) \\ (1+\lambda+\lambda^2) & (1+2\lambda^3) & (1+\lambda^2+\lambda^4) \\ (1+\lambda+\lambda^2) & (1+\lambda^2+\lambda^4) & (1+2\lambda^3) \end{bmatrix}$$

(IV.30)

Figure IV.9 shows how phasors are rotated with the λ operator. From the figure it may be seen that $(1 + \lambda + \lambda^2) = 0$, $(1 + 2\lambda^3) = 3$ and $(1 + \lambda^2 + \lambda^4) = 0$.

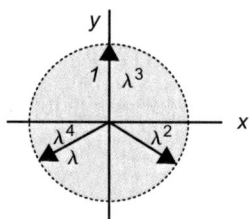

Figure IV.9 λ operator

From (IV.30):

$$\begin{bmatrix} 1 & 1 & 1 \\ 1 & \lambda^2 & \lambda \\ 1 & \lambda & \lambda^2 \end{bmatrix} \times \begin{bmatrix} 1 & 1 & 1 \\ 1 & \lambda & \lambda^2 \\ 1 & \lambda^2 & \lambda \end{bmatrix} = 3 \times \begin{bmatrix} 1 & 0 & 0 \\ 0 & 1 & 0 \\ 0 & 0 & 1 \end{bmatrix}$$

(IV.31)

Therefore from matrix algebra:[2]

$$\begin{bmatrix} 1 & 1 & 1 \\ 1 & \lambda^2 & \lambda \\ 1 & \lambda & \lambda^2 \end{bmatrix}^{-1} = \frac{1}{3} \begin{bmatrix} 1 & 1 & 1 \\ 1 & \lambda & \lambda^2 \\ 1 & \lambda^2 & \lambda \end{bmatrix}$$

(IV.32)

Thus from (IV.29):

$$\begin{bmatrix} V_{A0} \\ V_{A1} \\ V_{A2} \end{bmatrix} = \frac{1}{3} \begin{bmatrix} 1 & 1 & 1 \\ 1 & \lambda & \lambda^2 \\ 1 & \lambda^2 & \lambda \end{bmatrix} \begin{bmatrix} V_A \\ V_B \\ V_C \end{bmatrix}$$

(IV.33)

IV.5   Problems

1.   A single-phase transformer has a primary/secondary turns ratio of 0.5. The total resistance and reactance referred to the secondary side are 2.5 Ω and 10 Ω

[2] $[A] \times [B] = 3[I]$, $[A] \times [B]/[A] = 3[I]/[A]$, $[B] = 3[A]^{-1}$, $[A]^{-1} = 1/3[B]$

respectively. If its secondary output voltage is 120 V when the output current is 10 A at 0.8 lagging power factor, calculate the input voltage.

(*Answers*: 420.6 V)

2. For the circuit shown in Figure IV.Q2, using per unit representation on 25 MVA base calculate the current in the line and the internal generator voltage in pu. Assume that the voltage at the infinite busbar is held at 1 pu.

(*Answers*: $I = 0.2 - j0.04$ pu; $E_F = 1.13 \angle 14.1°$)

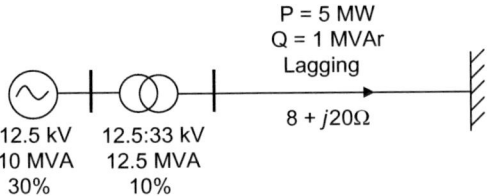

Figure IV.Q2

3. Convert the part of the distribution network shown in Figure IV.Q3 into a per-unit equivalent circuit on a 5 MVA base.

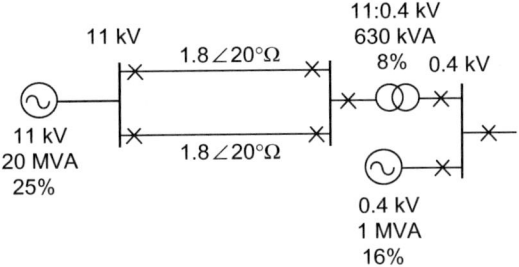

Figure IV.Q3

IV.6 Further reading

1. Chapmen S.J. *Electrical Machinery Fundamentals*. McGraw-Hill; 2005.
2. Grainger J.J., Stevenson W.D. *Power System Analysis*. McGraw-Hill; 1994.
3. Weedy B.M., Cory B.J. *Electric Power Systems*. John Wiley; 2004.

Glossary

Active network management: Active control of distribution networks in response to network conditions, including generator control and network switching (reconfiguration).

Anti-islanding protection: Electrical protection to detect if a section of the distribution network has become isolated or 'islanded'. The distributed generator is tripped to prevent continuous operation of the autonomous power 'island'. Effectively synonymous with loss-of-mains protection.

Asynchronous generator: Synonymous with induction generator. The rotor runs at a slightly faster rotational speed than the stator field. The construction of an asynchronous (or induction) generator is very similar to that of an induction motor of similar rating.

Auto-reclose: The automatic reclosure of a circuit breaker after it has opened when a fault has been detected.

Biomass: Biological material used as fuel for conversion to heat or electrical energy.

Capacity credit: Synonymous with capacity value. The capacity of the incumbent conventional generation that a distributed generator can displace, expressed as a percentage of the distributed generator capacity.

Capacity factor: Synonymous with load factor. The energy produced by a generator (usually measured over a year) divided by the energy that would have been produced if the generator were operating constantly at its rated output.

CCGT: Combined cycle gas turbine.

CHP, combined heat and power: The simultaneous production of heat and electrical energy. Synonymous with cogeneration and total energy. Industrial combined heat and power systems typically produce hot water or steam and electrical power for use within the host site. The heat output from combined heat and power plants may also be used for district heating.

Dispatched: Generating plant that is under central control and so 'dispatched' or controlled by the power system operator.

Distributed generation: Generation that is connected to the distribution network. Synonymous with dispersed and embedded generation, terms that are now falling into disuse.

DFIG, doubly fed induction generator: Variable speed generator that uses a wound rotor induction generator with back/back voltage source converters in the rotor circuit. Widely used in wind turbines to give variable speed operation of around ±30% of synchronous speed.

DUoS, distribution use of system charge: Charge made to generation and load customers for the use of the distribution network to transport electrical energy.

EENS: Expected energy not supplied.

Embedded generation: Synonymous with distributed generation.

Fault level: The fault level at a given point in a power network is the product of the three-phase fault current and the pre-fault voltage. It is a measure of the magnitude of the fault current that would result from a balanced three-phase fault at that point. The fault level is higher for networks that are more heavily meshed and increases as the point considered moves closer to generators. Synonymous with short-circuit level.

Fit-and-forget: The philosophy of planning distribution networks so that they can accommodate any anticipated combination of loads and distributed generation without active control of the network.

Flicker: Used to describe high frequency (up to 10 Hz) variations in network voltage magnitude which may give rise to noticeable changes in light intensity or 'flicker' of incandescent lamps.

Fuel cells: An electrolyte cell supplied continuously with chemical material, stored outside the cell, which provides the chemical energy for conversion to electrical energy.

FPC, full power converter: Variable speed system whereby all the power from the generator is rectified to DC and then inverted to AC through two voltage source converters rated at the full output of the generator. It allows variable speed operation over a wide speed range and full control of real and reactive power exported to the network.

Generation reserve: Generation reserve is used to balance generation and demand following the unexpected breakdown of plant or after a demand or renewable energy output forecast error.

Geothermal plant: Generation plant that uses heat from the earth as its input.

Grid supply point: A grid supply point is a substation where the transmission network is connected to the sub-transmission or distribution network. In England, this connection is usually made by transformers that step the voltage down from 400 kV or 275 kV to 132 kV.

Harmonic distortion: Distortion of the network voltage or current from a true sinusoid.

Induction generator: Synonymous with asynchronous generator.

Intertie: A line or group of lines that connect two power systems that are operated by different companies.

LOEE: Loss of energy expectation, i.e. the expected energy that will not be supplied due to those occasions when the load exceeds the available generation. It encompasses the severity of the deficiencies as well as their likelihood.

LOLE: Loss of load expectation, i.e. the average number of days on which the daily peak load is expected to exceed the available generating capacity. Alternatively it may be the average number of hours during which the load is expected to exceed the available capacity. It defines the likelihood of a deficiency but not the severity, nor the duration.

LOLP: Loss of load probability, i.e. the probability that the load will exceed the available generation capacity. It defines the likelihood of a deficiency but not the severity.

Load factor: Synonymous with capacity factor. The energy produced by a generator (usually measured over a year) divided by the energy that would have been produced if the generator were operating constantly at its rated output.

Loss-of-mains protection: Electrical protection applied to an embedded generator to detect loss of connection to the main power system. Synonymous with anti-islanding protection.

Neutral grounding: The connection of the neutral point of a three-phase power system to ground (or earth).

Neutral voltage displacement: Electrical protection relay technique used to measure the displacement of the neutral point of a section of the power system. Used particularly to detect earth faults on networks supplied from a delta-connected transformer winding.

OCGT: Open cycle gas turbine.

Permanent outage: An outage associated with damage faults that require the failed component to be repaired or replaced.

Photovoltaic: The physical effect by which light is converted directly into electrical energy.

Power: Electrical power is the product of a current and a voltage. In AC circuits, this product is called apparent power. The angle difference between the current and voltage waveforms is called the phase angle and the cosine of this phase angle is called the power factor. The real power is the electrical power that can be transformed into another form of energy. It is equal to the product of the apparent power and the power factor. Reactive power is not transformed into useful work and is a mathematical construct used to represent the oscillation of power between inductive and capacitive elements in the network. It is equal to the product of the apparent power and the sine of the phase angle. A purely active load (i.e. a load such as an electric heater that consumes only real power) has a power factor of 1.0. Practical loads usually have a power factor smaller than 1.0. A load with a power factor of 0.0 would be purely reactive and would not transform any electrical energy into practical work.

Quality of supply: Perfect 'quality of supply' in the most general sense means an undistorted waveform without any interruptions of any duration. Some organisations however associate quality only with waveform distortions, e.g. harmonics, voltage sags, etc. Others associate it with short and/or long interruptions of supply.

Radial distribution feeder: A distribution feeder (underground cable or overhead line) that is connected to one supply point only.

Rate of change of frequency (ROCOF): An electrical protection technique based on the rate of change of electrical frequency used in loss-of-mains or anti-islanding relays.

Reliability: This is an inherent attribute of a system, which is characterised by a specific group of measures that describe how well the system performs its basic function of providing customers with a supply of energy on demand and without interruption.

Scheduled maintenance outage: An outage that is planned in advance in order to perform preventive maintenance.

Stability: Under normal circumstances, the power network is capable of handling the flow of power from generators to loads. Following unpredictable events, such as faults or failures, the network may lose its ability to transfer this power. Under such circumstances, the power system is said to have lost its stability. Stability can be lost either through transient instability or voltage instability. Transient instability means that, following a fault, the rotor angle of one or more generators increases uncontrollably with respect to the angle of the other generators. Voltage instability means that, due to a lack of reactive power generation near the loads, the voltage in parts of the system collapses uncontrollably.

Stirling engine: An external combustion engine based on the Stirling thermodynamic cycle. Stirling engines are of considerable potential significance for small-scale CHP schemes.

Symmetrical components: Symmetrical components are a way of describing the voltages and currents in a three-phase system that is particularly useful when studying unbalanced conditions such as some kinds of faults.

Synchronous generator: A generator whose rotor operates in synchronism with its stator field. In its usual construction a synchronous generator allows independent control of real and reactive output power.

Temporary outage: An outage associated with transient faults, which is restored by manual switching, fuse replacement or similarly lengthy restoration times.

Transient outage: An outage associated with transient faults, which is restored by automatic switching.

Transient voltage variations: Rapid (greater than 0.5 Hz) changes in the magnitude of the network voltage. May be repetitive or may refer to a single event.

VSC, voltage source converter: Power electronic converter that uses transistor switches, which can be turned on and off, to create a variable frequency and magnitude voltage source. This is connected to the network through a coupling reactor. VSCs allow independent control of real and reactive power.

Voltage vector shift: A protective relay technique based on measuring the change or shift of the angle of the system voltage vector. The technique is used in loss-of-mains relays.

Index

Page numbers followed by *f* and *t* indicate figures and tables, respectively.